ADVANCED ENGINE PERFORMANCE SPECIALIST TEST

Automobile Technicians Refresher Course

Contents

Chapter One

GENERAL POWERTRAIN DIAGNOSIS

To effectively diagnose and repair driveability, engine performance, and emissions failure problems, a technician must be organized and use a systematic and logical approach. Locating the source of a malfunction is a process of elimination that begins with general area checks and leads to specific circuit and component tests. Skipping steps and taking short cuts often ends with inconclusive results. In this chapter we discuss the preliminary inspections and evaluations performed as the first step in troubleshooting. Always begin your diagnosis with an overall vehicle evaluation and basic system inspections.

VEHICLE EVALUATION AND BASIC INSPECTION

The electronic control system does not monitor or control all functions of the engine. Mechanical systems and components, such as vacuum and coolant lines, wiring and connections, basic fuel delivery, ignition operation, and cooling system condition must all be inspected before you begin troubleshooting. Also check for body damage, mechanical damage, tampering, and newly installed accessories that might compromise system operation. Begin by:

- Verifying the complaint
- Inspecting the vehicle
- Checking the engine subsystems

Conducting a thorough preliminary evaluation will often reveal the cause of, or conditions that contribute to, common driveability complaints.

Verifying The Complaint

Knowing the symptoms of a particular problem will help organize your approach. The owner of the vehicle is the best source for this information. Get the customer to describe the problem in as much detail as possible. It is important to know what the problem is, conditions under which it occurs, and the severity of the symptoms. Typically, you will want to know:

- If the problem exists all the time or some of the time, whether it occurs regularly or at random, and if it is happening now.
- If certain conditions, such as accelerating or climbing hills, contribute to the problem.
- If the problem occurs at certain temperatures, such as on cold starts or after a hot soak.
- What all the symptoms are, including noises, vibrations, smells, performance, or any combination.
- Whether the problem has occurred before, and what was done to repair it.
- When was the vehicle last serviced and what work was performed.

Be aware, the way a customer describes a problem may be very different than the way you would describe it. In some cases, it can be advantageous to have the customer accompany you on a test drive to pinpoint symptoms when they occur.

Vehicle Inspection

Inspect for obvious faults and try to eliminate simple problems first. Check for loose or broken wires, connectors, and vacuum hoses. Inspect for oil and coolant leaks, and for worn drive belts and hoses. Compare vacuum hose routing to the underhood emissions decal to ensure that everything is properly connected, figure 1-1. Also check for mechanical and electrical tampering and collision damage.

A late-model electronic control system is carefully calibrated to meet the needs of a specific vehicle in stock configuration. Control tolerances are very narrow, and even minor modifications can compromise system accuracy and performance. Simply installing non-stock wheels and tires can upset the balance between vehicle speed and engine speed leading to an increase in emissions.

Practically any modifications, whether minor or all out high-performance, can cause I/M 240 test failures. Because I/M 240 testing is done on a dynamometer, emissions are measured as the vehicle operates through all gears. The PCM may be unable to maintain stable engine combustion efficiency if transmission shifts during the test sequence fail to match those of the programmed factory schedule. Any change to the driveline, including installing a shift kit or changing the

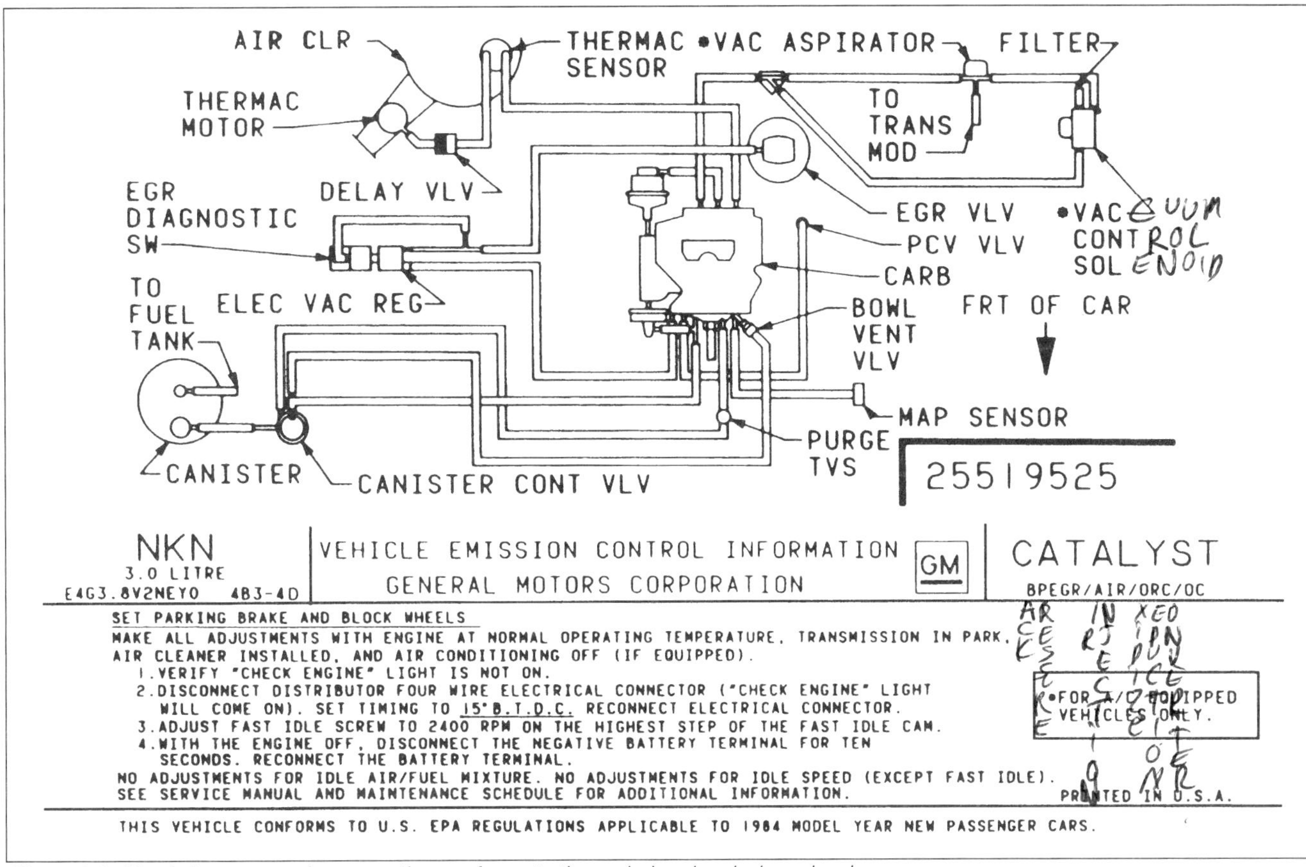

Figure 1-1. Check that vacuum hoses routing conforms to the underhood emissions decal.

final drive gear ratio can effect tailpipe emissions, and may result in a test failure. Exhaust analysis is detailed in Chapter 6 of this book.

In addition to driveline modifications, look for any sign of tampering. The electrical system is vulnerable, accessories that are not wired properly can cause problems. Aftermarket installers of stereos and alarms are notorious for tapping into PCM controlled circuits to power their accessories. Keep these points in mind during your preliminary inspection. Any non-stock component should raise a warning flag. Repair any obvious problems now, before you continue with your diagnosis.

In order to run, any engine requires three elements; fuel, compression and spark. Perform the following checks to verify all three are present:

- **Fuel** — Use a gauge to check fuel pump pressure, figure 1-2. Also check fuel pump delivery volume where applicable.
- **Compression** — Do a compression test or a cylinder power balance test, if possible, to check basic engine operation.
- **Spark** — Use a spark tester to check for coil secondary voltage.

Once the these basic requirements are met, continue with more specific checks of the subsystems.

Check The Engine Subsystems

Most mechanical parts and many electrical subsystems are not monitored by the engine control system. Failures here can cause driveability problems without setting codes and cannot be detected on the serial data stream, or with an oscilloscope. At this point, we list checks that should be made. Specific procedures

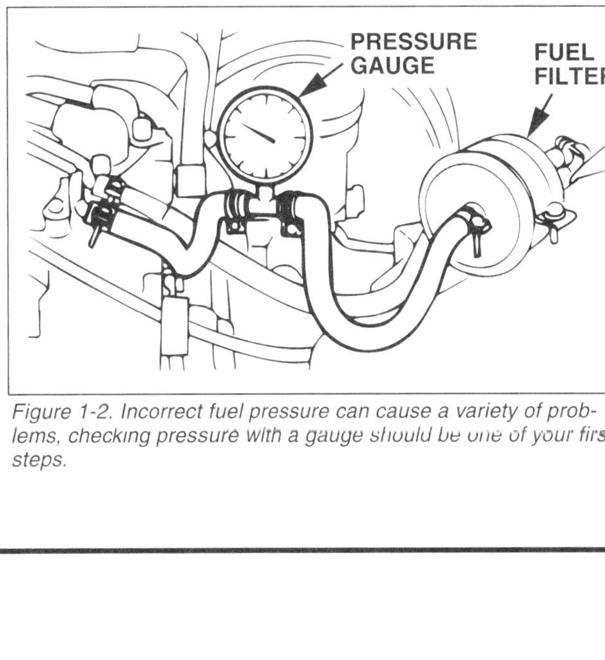

Figure 1-2. Incorrect fuel pressure can cause a variety of problems, checking pressure with a gauge should be one of your first steps.

for performing tests are covered in detail in the subsequent chapters of this book. Check the following:

Cooling system

Be sure the radiator is properly filled with the correct mixture of antifreeze/coolant and water. Look for rust and other contamination in the coolant. Check the operation of the thermostat and cooling fans. Cooling system problems can fool the engine control system into compensating for unrealistic operating conditions. Malfunctions that lower engine operating temperatures, such as a stuck open thermostat, can cause the control system to command a rich fuel mixture. The PCM may compensate for higher than normal temperatures by altering fuel delivery, spark timing, and idle speed. In addition, converter clutch operation, canister purging, and exhaust gas recirculation may be suspended until engine temperature falls into normal range. An engine running outside its normal operating temperature range, whether too hot or too cold, can cause an emissions test failure.

Battery

The battery must be fully charged and provide adequate cranking current. Minimum battery open circuit voltage is 12.4 volts for proper cranking, figure 1-3.

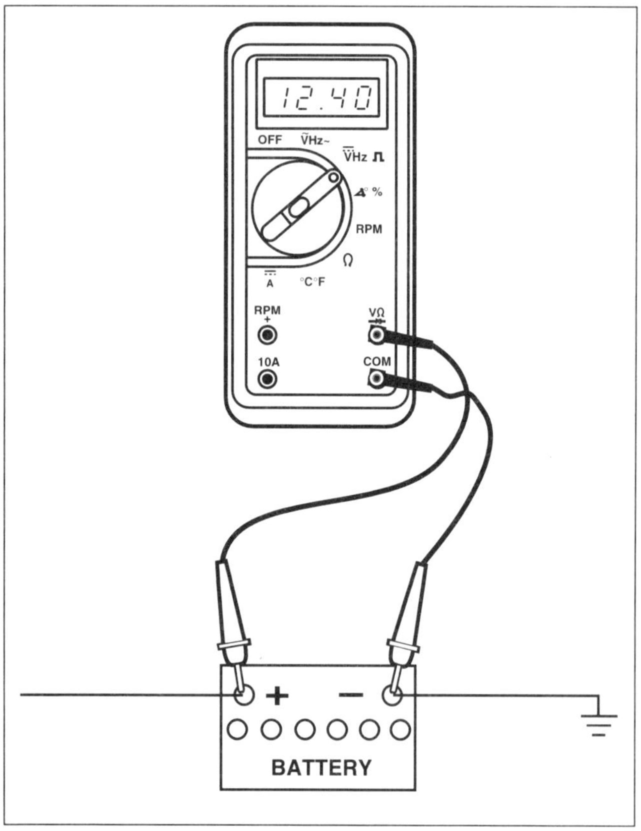

Figure 1-3. Open circuit battery voltage must be at least 12.4 volts to ensure proper control system operation.

Most electronic control systems require at least 10 volts to work properly throughout a diagnostic sequence. If battery voltage is low, charge or replace the battery before continuing.

Cranking operation

The engine should crank at a normal speed without drawing excess battery current. For a no-start problem, remove the distributor cap and be sure that the rotor turns.

Some systems contain an override that prevents cranking if certain conditions are not met. These include safety switches that prevent cranking if an automatic transmission shift lever is not in Park or Neutral, and clutch switches for manual transmission vehicles that require you to depress the clutch before cranking. If these switches are faulty, they may prevent cranking even if conditions *are* met.

Mechanical condition

Do a compression test or a cylinder power balance test, if possible, to check basic engine operation. Use a vacuum gauge to check manifold vacuum with the engine both cranking and running. This can often help you spot a restricted exhaust system. Check the oil level and condition, and inspect the PCV system. The adaptive strategy on many control systems will mask the driveability symptoms of some marginal mechanical problems. However, a mechanical problem results in inefficient combustion that can be detected through exhaust gas analysis, see Chapter 6 for more detail.

Fuel quality

If you suspect fuel contamination, use a gasoline-checking kit to test the quality. Low-grade fuel not only hampers engine performance, it can also effect control system strategy, increase emissions, and lead to mechanical failures.

Ignition system

Do a basic oscilloscope check of the ignition system to evaluate its overall operation. Chapter 3 of this book covers ignition system diagnosis is detail.

Ignition timing

Although spark advance is controlled by the PCM on most late-model engines, base timing must be set manually on all but those engines with distributorless ignition systems (DIS), figure 1-4. Procedures for checking and adjusting base timing vary, so consult a service manual, or the emissions label, and follow the instructions exactly.

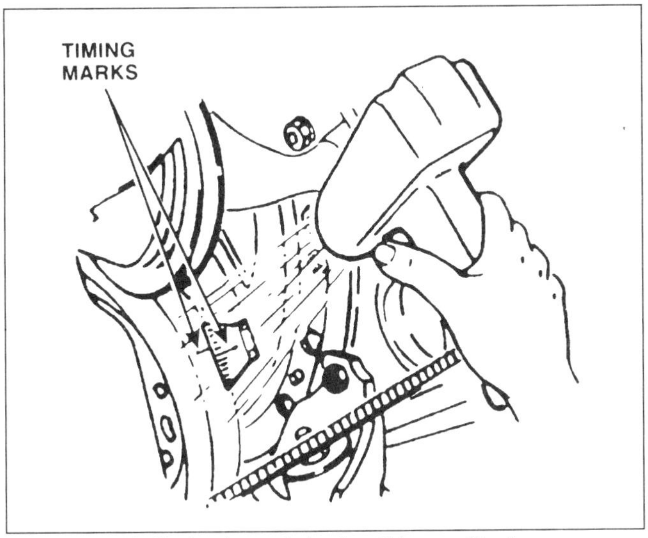

Figure 1-4. Verify that base timing is within specification.

Electronically-controlled spark advance

On most engines, ignition timing should advance as you accelerate the engine in park or neutral. On a few engines, the PCM prevents spark advance unless the transmission is in a forward gear, or until the vehicle reaches a certain speed. If you do not get advance, check and follow the manufacturer's instructions.

Air intake system

An electronically-controlled engine can suffer problems from not enough intake airflow or from too much intake airflow. A restricted air filter can cause a rich exhaust condition. This is a common cause of a rich exhaust on carbureted engines. However, most electronically controlled fuel-injected systems can compensate to a greater or lesser degree, depending on the system.

Unmetered air and vacuum leaks are common causes of several driveability problems. These problems include a lean exhaust, rough idle, or backfiring on some engines.

Idle speed

Idle speed is controlled by the PCM on most late-model engines, check it to be sure the computer is controlling it correctly. Vacuum leaks can cause a rough or high idle that is beyond the control range of the PCM. Dirty fuel injectors can also cause a rough idle. The PCM should maintain a steady idle as you switch the air conditioner on and off. Check the throttle position (TP) sensor signal as part of idle speed test and adjustment, figure 1-5.

Minimum idle speed for fuel-injected engines

A specification for the minimum speed at which a fuel-injected engine will idle is provided by the manufacturer. This may also be called the minimum air rate. The procedures for checking and adjusting minimum idle speed vary, so check a service manual and follow the instructions exactly. An improperly adjusted minimum idle speed can result in an incorrect signal from the throttle position sensor and cause a driveability problem.

Throttle body coking is another common problem that affects the minimum idle speed on engines with port-fuel injection. Coking is a build up of carbon deposits on the throttle plates and the idle air control valve. These deposits can prevent the throttle and idle valve from seating. The result can be a number of driveability problems, including an irregular idle speed, stalling on deceleration, and an off-idle hesitation. Except in the most severe cases, throttle body coking will not set a trouble code.

Exhaust emissions

Checking the tailpipe with an infrared analyzer is often an important early step in solving a driveability problem, and is essential if the vehicle has failed an emissions test. At the minimum, check exhaust HC and CO emissions both at idle and at no-load cruising speed (2000 to 2500 rpm). For interpreting the results exhaust gas sampling, see Chapter 6 of this book.

After completing your basic inspections, and eliminating any problems, your next step is to prepare the engine for testing.

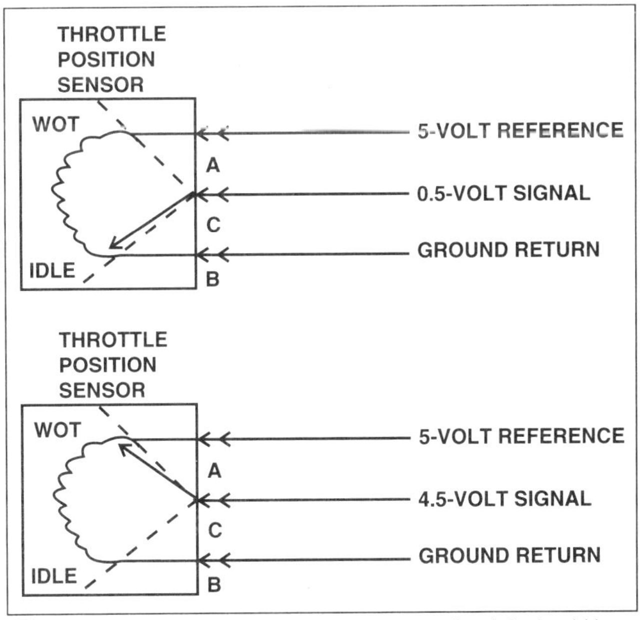

Figure 1-5. Check the throttle position sensor signal, it should be low with the throttle closed and high with the throttle wide open.

PREPARING THE ENGINE FOR TESTING

To test an engine control system thoroughly, the engine should be warmed to normal operating temperature. However, some initial tests can be performed on a cold engine. For some procedures, you must start with a cold engine. Troubleshooting a no-start problem is a good example. Other instances would include problems that only occur when the engine is cold, or operating in open loop. In many cases, you will want to check engine operation immediately after a cold start, then watch the data readings as the engine warms and goes into closed loop.

Many driveability complaints and system trouble codes will relate to warm, closed-loop operation. One basic test of any engine control system is to verify that it goes into closed-loop operation. As a minimum, the engine should be run for 2 to 3 minutes above 2000 rpm. This allows the oxygen sensor, catalytic converter, and coolant temperature to come up to operating range and permits the system to enter closed loop. Even when warm, many systems require several minutes of engine operation to let the open-loop timer in the PCM expire before they will enter closed loop. To warm the engine completely, you may have to run it for 10 to 15 minutes or drive the vehicle for several miles.

Preparation for Emission Testing

Most emission test programs require that the engine be conditioned in a specific way before testing. The California BAR 90 emission test program (also followed by several other state emission programs) calls this procedure "preconditioning" because it must performed as a pretest requirement. The basic conditioning procedure is as follows:

- With the parking brake set and the transmission in park or neutral, start and run the engine at 2200 to 2800 rpm for 3 minutes. Allow the engine return to idle, then run at idle for 10 seconds. Then, immediately proceed with the emission test without switching the engine off.

This warmup conditioning procedure ensures the following operating conditions, figure 1-6:

- Cold start enrichment is completed, and the engine is not running with an overly rich mixture.
- The catalyst is warmed to efficient operating temperature, minimum temperature is about 450°F (232°C).
- The exhaust oxygen sensor (O2S) is warmed to at least 572°F (300°C) and sends a varying feedback voltage signal to the PCM.
- Engine coolant is at, or near, warmed up operating temperature.

PCM
- STARTUP TIMER EXPIRED (2 MINUTES)
- COLD START ENRICHMENT COMPLETE
- RELIABLE O2S SIGNAL RECEIVED
- OPERATING IN CLOSED LOOP

ENGINE COOLANT
- TEMPERATURE OVER 150°F(66°C)

OXYGEN SENSOR
- TEMPERATURE OVER 600°F(315°C)
- SENDING A VARIABLE VOLTAGE SIGNAL

CATALYTIC CONVERTER
- TEMPERATURE OVER 450°F(232°C)

EVAP CANISTER
- INITIAL PURGING COMPLETE

Figure 1-6. Preconditioning ensures that the system is in closed-loop, all prerequisites have been met, and the vehicle is ready for testing.

- All other sensor signals that are sensitive to time or temperature are within normal operating range.
- The internal timer of the PCM has expired.
- Evaporative canister purging, which usually is heavy after start-up, is completed.

Prior to testing, OBD-II compliant vehicles must be operated through a specific drive cycle in order to perform all of the emission control monitor checks. These monitors test the integrity of emission systems components at various speeds, loads, and temperatures during normal driving conditions. The system must complete, and pass, all of the monitors before the vehicle can be tested for certification. Exact driving conditions to complete the monitors will vary slightly between vehicles. Figure 1-7 shows a typical OBD-II drive cycle.

Even if emission testing is not the primary job to be performed on a vehicle, these warmup conditioning procedures are usually an important part of control system troubleshooting.

CONTROL SYSTEM DIAGNOSIS

When evaluating the control system, it is important to test from the general to the specific. If you immediately look at the controlled circuits for the cause of a problem, you can overlook other possible causes in the fuel, ignition, and emission systems. The basic inspection described in the previous section begins with general checks, or area tests. Testing the electronic control system should also begin with general area tests before moving on to pinpoint tests.

Late-model control systems have the ability to store an electronically encoded message whenever a fault occurs on one of the monitored circuits. These messages, or diagnostic trouble codes, provide a valuable diagnostic aid.

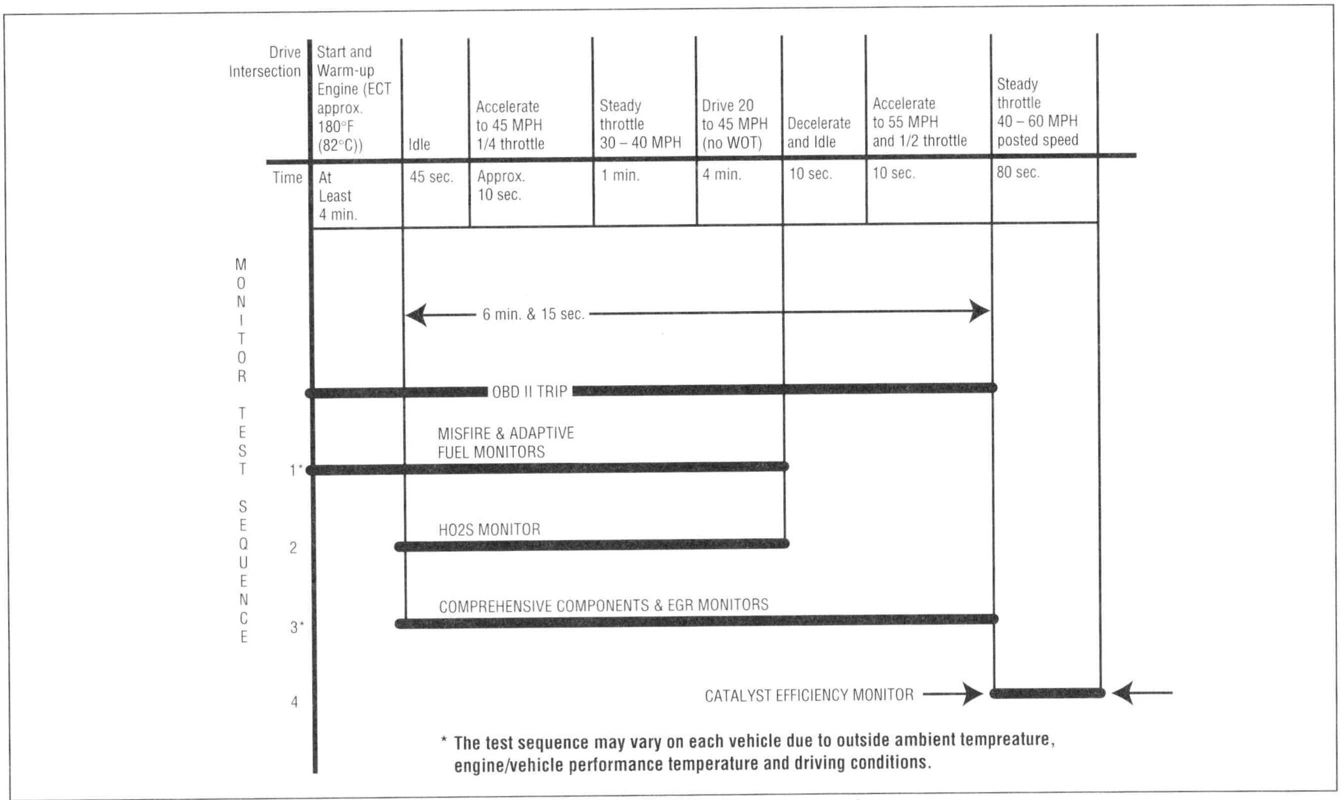

Figure 1-7. An OBD-II drive cycle guarantees that all emission control monitoring is complete.

Diagnostic Trouble Codes

Diagnostic Trouble Codes (DTC's) monitor important PCM circuits anytime the ignition switch is on or the engine is running. If any of these monitored circuits is open, grounded, or has an out of range condition, the resulting high or low voltage, or frequency value on the affected circuit causes the PCM to store an associated fault code. A DTC describes problems about a circuit, not a particular component failure within the circuit.

All late-model cars and light trucks are required to have a malfunction indicator lamp (MIL) on the instrument panel that lights to indicate major system problems, figure 1-8. The lamp may be labeled ENGINE, CHECK ENGINE, SERVICE ENGINE SOON, or some term that indicates a system problem. The MIL normally comes on if there is a failure in a circuit that is being monitored by the PCM. If the MIL remains illuminated because of a circuit failure, the PCM usually goes into a back-up, limp-home, or failure effects management mode. The MIL remains on as long as the fault causing the problem is present. If the circuit condition returns to normal, the MIL goes out and the vehicle returns to normal operation. In most cases, the computer sets a diagnostic trouble code (DTC) in long term memory to identify the problem circuit or circuits.

Figure 1-8. The MIL must light when the ignition key is switched on without starting the engine.

Parameters for setting a DTC:

- Signal is not furnished (open circuit)
- Signal is high
- Signal is improbable
- Signal is out of range for extended time period
- Signal disagrees with values from other sensors
- Signal comes and goes (intermittent)
- Signal is erratic

OBD II trouble codes

OBD-II compliant vehicles, as well as the ASE L1 composite vehicle, use industry standard 5-character codes, figure 1-9.

There are four letters currently assigned for system recognition to be used as the first character of the DTC: These are "B" for Body, "C" for Chassis, "P" for Powertrain, and "U" for Undefined. Undefined codes are reserved for future assignment by SAE, and definitions for body and chassis codes are not finalized yet.

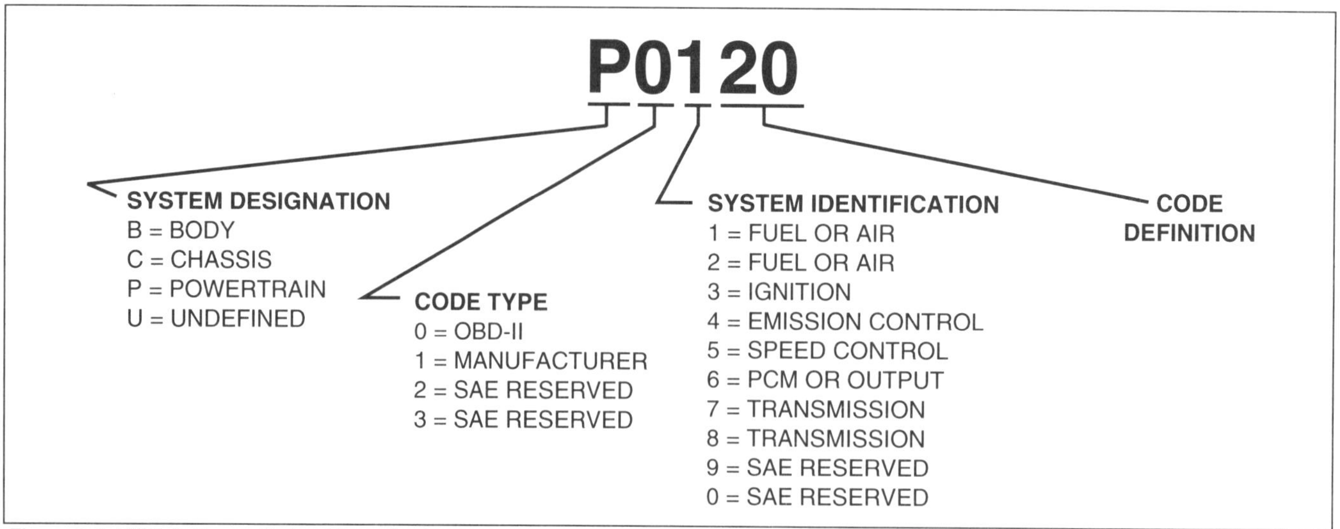

Figure 1-9. OBD II diagnostic trouble code structure.

The second character of the DTC will be either a "0", "1", "2", or "3". A zero in this position indicates a malfunction that is defined by OBD-II; the number one is a code defined by the vehicle manufacturer. The numerals two and three are designated for future use and both are reserved for SAE assignment in powertrain codes. With body and chassis codes, the digit two is manufacturer controlled, and three is reserved for SAE.

The third character of a powertrain DTC indicates the system where the fault occurred. Both "1" and "2" are used to designate fuel or air metering problems. A "3" in this position indicates an ignition malfunction or engine misfire. Number "4" is assigned to the auxiliary emission control system. Problems in the vehicle or idle speed control system will set a number "5" in the DTC. Number "6" is used for computer or output circuit faults. Transmission control problems are indicated by either a "7" or "8" while "9" and "0" are reserved for future use.

The remaining two digits of the DTC tell you the exact condition that triggered the code. For the L1 test, ASE has assigned 17 five-character, OBD-II style codes for the composite vehicle. They are defined in the *Composite Vehicle Preparation/Reference Booklet* that ASE sends out when you enroll for the L1 test.

MIL Operation

The MIL illuminates when the ignition switch is first turned on as a bulb check. The PCM turns the MIL off as soon as the engine starts and it receives a distributor or crankshaft reference signal. If the engine is running and the MIL remains on, the PCM has detected a problem or cannot turn off the light due to an external or internal circuit failure.

Checking MIL function

If the MIL does not light with the key on, the PCM probably will not go into the diagnostic mode. The problem may be as simple as a burned out bulb, it may be a faulty circuit, or could be an internal PCM fault. Begin by checking the bulb. Then test the lamp circuit for correct power and ground.

If the MIL lights steadily with the engine running, it indicates that a system problem exists and a diagnostic trouble code has been recorded. Not all system faults will light the MIL. With some faults, the MIL may flash briefly when the code is recorded, then go out. Other problems will set a trouble code in memory but will not light the MIL at all. Code setting conditions vary by manufacturer, year, and model. You must have accurate service information for the vehicle being tested.

Check voltage at the data link connector (DLC)

If the MIL does not light with the key on or if the PCM will not go into a diagnostic mode, you will have to make some voltage checks at the test connector. Most diagnostic connectors have a ground terminal that is used for one or more test modes. On an OBD-II 16-pin connector, the chassis ground is at pin 4, and the signal ground is at pin 5. Other test connectors do not have standard pin assignments, so you will need to locate this information in the appropriate service manual.

Check for continuity at the ground terminal with a voltmeter or ohmmeter, figure 1-10. High resistance or an open circuit can keep the PCM out of the self-test mode and may provide a clue to other system problems. Various other terminals on the diagnostic connector may have other levels of voltage applied to them at different times. Some may have battery (system)

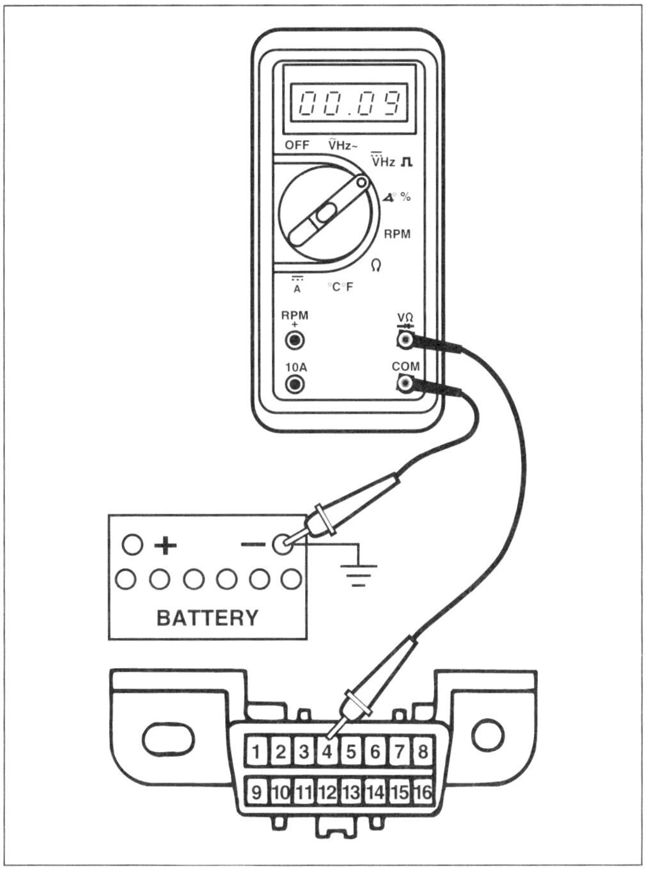

Figure 1-10. As a continuity check, voltage drop at the DLC ground terminal should be less than 0.1 volt.

voltage present under certain conditions. Others may have 5-volts, 7-volts, or a variable voltage applied for specific test conditions.

Check for Codes in Memory

Once all of the preliminary checks have been made, check for any diagnostic trouble codes stored in memory. To access PCM memory, the system must be placed in diagnostic mode. Procedures for entering diagnostics varies by manufacturer and system. Most systems will display codes as a series of flashes on the MIL, or on the display screen of a scan tool. The following paragraphs outline the common features and principles of trouble code diagnosis.

When the PCM recognizes a signal or condition that is absent or out of range, a diagnostic trouble code (DTC) is stored in memory. A DTC can indicate a problem in a particular circuit or subsystem, however, does not pinpoint the exact cause of the problem. Checking codes can direct you to where to start your troubleshooting. In general, trouble codes fall into one of two categories: those that indicate a fault that exists at the time of testing, and those that indicate a fault that occurred in the past but is no longer present. Although manufacturers use different terminology to describe the types of codes, we will refer to codes for faults present during testing as hard codes, and codes for faults not present during testing as soft codes.

Hard Codes

A hard code indicates a failure that is present at the time of testing and remains permanently in the system until the cause has been repaired. If you turn off the ignition and clear the codes, a hard code will reappear immediately or within a few minutes because the problem that set the code still exists in the system. Hard codes indicate a full-time problem that requires immediate attention. These are top priority problems, and must be repaired first, before any other code repairs are made. Hard codes are generally easy to diagnose because the conditions that caused the code to set are present at the time of testing.

A hard code enables you to go right to a certain area, or areas, and begin pinpoint testing to isolate the defective component. Most diagnostic charts are designed to troubleshoot hard codes, figure 1-11. The procedures assume that the problem is present at the time of testing.

Soft Codes

A soft code indicates an intermittent problem: one that comes and goes. The PCM records soft codes when they occur, then maintains a record of the fault in memory. A soft code represents a problem that occurred some time in the past, before testing, but is not present now. The problem, and the DTC, may not reappear if you clear the codes from memory and retest the system. The conditions to set the code may happen only at a certain speed, temperature, or other condition that you cannot recreate in the shop.

Because soft codes indicate intermittent problems, diagnostic charts usually do not isolate the problem immediately. Often you will need to follow special intermittent, or symptom diagnostic test procedures to troubleshoot soft codes accurately. To find the problems that cause soft codes, do not open or disconnect electrical connectors until you have checked them in normal operation or performed a wiggle test. Disconnecting and reconnecting a connector may temporarily solve a problem without revealing the basic cause.

Checking codes at this point is a preliminary step, it will not provide a definitive answer as to the source of the problem. Some codes you find now, especially soft codes, may be erased from PCM memory as you continue your diagnosis. Write them down for future reference. Then start and run the engine, if possible, to verify that the system is operational and the PCM is controlling engine functions.

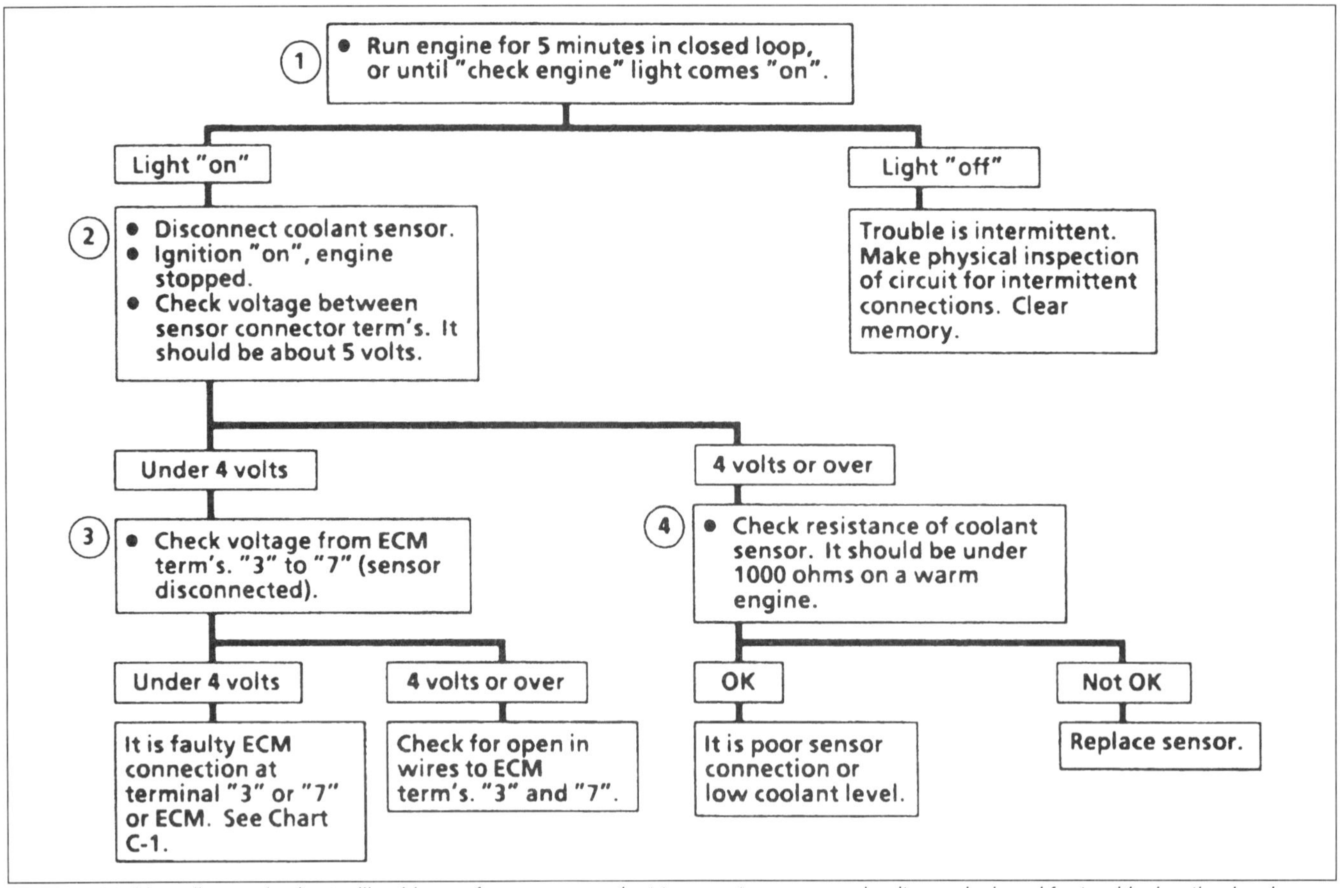

Figure 1-11. Most diagnostic charts, like this one for an open coolant temperature sensor circuit, are designed for troubleshooting hard codes only.

Checking Open-Loop And Closed-Loop Operation

All late-model engine control systems have two basic operating modes: open loop and closed loop. In open loop, the PCM ignores the exhaust oxygen sensor (O2S) and controls fuel metering based on engine speed, load, and temperature signals and its own program. In closed loop, the PCM responds to the rich and lean signals of the O2S as well as signals from other sensors. The PCM then controls fuel delivery to maintain the air-fuel ratio in relation to overall engine requirements. In most conditions, except high-power operation, the ratio is the 14.7:1 stoichiometric ratio.

A cold engine should start in open loop and go to closed loop as soon as the O2S is warm enough to send an accurate voltage signal to the PCM. Most systems also have a timer circuit that keeps the engine in open loop for several minutes after any start-up regardless of O2S and engine temperatures. In addition, some systems return to open loop at idle and during full-throttle acceleration. For accurate diagnosis, you must know the open-loop and closed-loop characteristics of the particular engine you are servicing. Along with checking trouble codes, verifying open-loop and closed-loop operation is a primary point in diagnosing most driveability problems. Figure 1-12 summarizes the loop control operating strategy of the composite vehicle used for the L1 Certification Test.

A scan tool can be used to determine if an engine is operating in open loop or closed loop. Most systems transmit a digital parameter that lets you know what strategy the PCM is operating in. When a loop control parameter is not available, simply monitor O2S feedback signal voltage. If O2S voltage fluctuates rapidly above and below the mid-point of its operating range (generally, 450 millivolts), it is sending a reliable feedback signal and the system should be in closed loop. If O2S voltage does not cross the mid-point consistently, the PCM considers the signal unreliable and maintains open-loop control.

Separating Hard and Soft Codes

For effective troubleshooting, you must determine if codes are hard or soft. Some systems specify which type of code you are dealing with on the scan tool display, but most do not. To distinguish between a hard and a soft code, record all codes, then clear the PCM memory. Codes can be cleared from all OBD-II, and

SFI SYSTEMS OPERATION STRATEGY, OPEN AND CLOSED LOOP CONTROL

OPEN LOOP MODE:

Engages on initial startup when engine speed exceeds 400 rpm. Fuel injector pulse width is calculated by the PCM based on readings from the:

- Throttle Position Sensor
- Engine Coolant Temperature Sensor
- Intake Air Temperature Sensor
- Manifold Absolute Pressure Sensor
- Engine Speed Sensor

CLOSED LOOP MODE:

Engages when the following conditions are met:

- The Oxygen Sensor is at operating tempreature (600°F/315°C) and generating a reliable signal
- Engine Coolant Temperature is above 150°F (66°C)
- Two minutes have elapsed sinse startup

ACCELERATION ENRICHMENT MODE:

During wide open throttle (over 80%) operation, the system temporarily enters open loop mode. Fuel injector pulse width is increased in proportion to the increase in manifold absolute pressure and the change in throttle position.

DECELERATION ENLEANMENT MODE:

During deceleration, the PCM overrides closed loop operation and decreases fuel injector pulse width. Injector pulse width is calculated by the PCM based on the decrease in manifold absolute pressure and the change in throttle position.

Figure 1-12. L1 composite vehicle loop control strategy.

most late-model vehicles using a scan tool. If scan tool code clearing is not available, you may have to momentarily disconnect the battery or remove the fuse for the PCM power circuit. Follow the procedure specified by the vehicle manufacturer.

After clearing the codes, drive the vehicle and watch for the MIL to light, an indication that the code has reset. If a code reappears immediately or soon, a hard fault is indicated. If the code does not reappear quickly, it was most likely a soft code and an intermittent problem is indicated. If you determine that a code is hard, go to the test or troubleshooting chart for that code number. If you determine that a code is soft, use intermittent diagnosis procedures to pinpoint the problem.

Trouble codes should be diagnosed and serviced in a basic order: hard codes first, followed by soft codes. Most vehicles transmit codes in numerical order from the lowest to the highest, and this is usually the order in which they should be serviced. Exceptions would be codes that indicate communications and internal PCM problems. These are priority situations that should be serviced first, before any other codes.

Driveability Symptom Tests

Many engine control system problems will produce clear and recognizable driveability symptoms but will not set a code. Typical driver complaints or symptoms include:

- Rough or high idle
- Surging or hesitation
- Stalls at idle or during deceleration
- Pinging, or detonation
- Poor fuel mileage

Test procedures or troubleshooting charts for these and other common symptoms are usually available in the factory service manuals. However, in order to use these tests, the symptom must occur or be present at the time of testing. Symptomatic tests are not designed for use on intermittent problems.

When using a symptomatic test, be sure to follow the one for the specific vehicle year, model, and engine. The tests are often different for cold and for warm engines.

Operating Range Tests

The signal from an analog sensor can drift out of range as the sensor ages or wears. Some sensors can develop an erratic signal, or dropout, at one point in the signal range. A loose or corroded ground connection for a sensor can also force the signal out of range, but not far enough to set a code.

These and similar problems can cause definite driveability problems without setting a code. You can check the operation of many sensors by using the operating range charts provided by the manufacturer. These charts list signal range specifications for voltage, resistance, frequency, or temperature that the sensor provides under varying conditions, figure 1-13.

Use a digital volt-ohmmeter (DVOM), a frequency counter, or other appropriate instrument to test the sensor signal. Take your initial readings at the sensor connector. Then, if necessary, at the main connector to the PCM. You can backprobe many sensor connectors or install jumper wires to provide connection points for your meter. You may need a breakout box or harness to check sensor signals at the main PCM connector. If possible, operate the sensor through its full range and check the signal at several points.

PCM Pin Voltage Charts

A PCM pin voltage chart identifies all the connector terminals at the main harness connector by number, circuit name, and the voltage levels that should be present under various conditions. Some circuits have different

Temperature F	Temperature C	Sensor Voltage
248	120	0.25
212	100	0.46
176	80	0.84
150	66	1.34
140	60	1.55
104	40	2.27
86	30	2.60
68	20	2.93
32	0	3.59
-4	-20	4.24
-40	-40	4.90

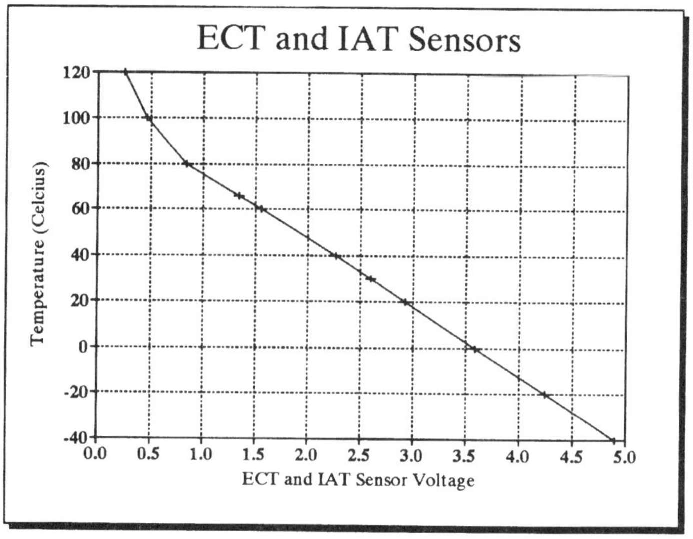

Note: Both sensors have the same temperature/voltage relationship

Vacuum at sea level [in. Hg]	Manifold Absolute Pressure [kPa]	Sensor Voltage
0	101.3	4.0
3	91.2	3.5
6	81.0	3.0
9	70.8	2.5
12	60.7	2.0
15	50.5	1.5
18	40.4	1.0
21	30.2	0.5
24	20.0	0.0

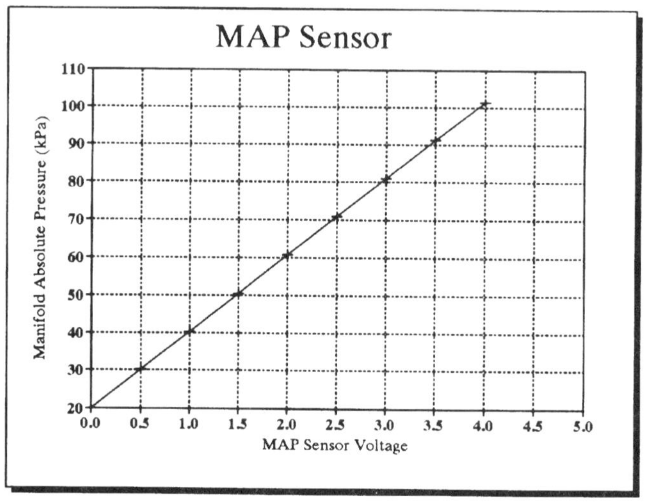

Figure 1-13. Operating range charts, like these for the L1 composite vehicle, allow you to make checks under variable conditions.

voltage specifications with the key on and the engine off, during cranking, and when the engine is running. Use the pin voltage charts to check input signals to the PCM and output signals from the PCM, figure 1-14. Checking signals at the PCM harness connector is closely related to sensor operating range tests.

Check Ground Integrity

Use a digital voltmeter to check the voltage drop across the main PCM ground connection, figure 1-15, and across the ground connection of any sensor that you think may be causing the problem. Low-resistance ground connections are critical for electronic control circuits.

With the ignition on, voltage drop across the ground connection for an electronic circuit should be 0.1 volt or less. The voltage drop across a high-resistance ground connection in series with a sensor circuit increases the signal voltage of the sensor. This ground resistance can offset the signal voltage enough to cause serious driveability problems. For example, on a throttle position sensor that operates on a 5-volt reference, a 0.5-volt drop across the ground connection equals a 10-percent error in throttle angle measurement.

IGN. "ON" (Voltage)	ENG RUN (Voltage)	OPEN CKT. (Voltage)	CIRCUIT	PIN	WIRE COLOR
0	14		FUEL PUMP RELAY	A1	DK GRN WHT
12	14	12	A/C CLUTCH CONTROL	A2	DK GRN/ YEL
12	14	12	CANISTER PURGE CONTROL	A3	DK GRN/ YEL
12	14	12	EGR CONTROL	A4	GRY
0	14	12	"CHECK ENGINE" CONTROL	A5	BRN/WHT
12	14	12	IGN - ECM FUSE	A6	PIN/BLK
0	0	.075	TCC CONTROL	A7	TAN/BLK
2-5	2-5		SERIAL DATA	A8	ORN
5	5	0	DIAG TERM	A9	WHT/BLK
		0	SPEED SENSOR SIGNAL	A10	BRN
		0	MAT	A11	DK GRN/ YEL
0	0	0	GRN'D	A12	BLK/WHT

WIRE COLOR	PIN	CIRCUIT	IGN. "ON" (Voltage)	ENG. RUN (Voltage)	OPEN CRT. (Voltage)
ORN	B1	BATT 12 VOLTS	12	14	12
TAN/WHT	B2	FUEL PUMP SIGNAL	0	14	0
BLK/RED	B3	EST REF LOW	0	0	0
WHT	B4	EST CONTROL	0	1.3	0
PPL/WHT	B5	EST REF HI	0	1.6	0
BRN/WHT	B6	MASS AIRFLOW SENSOR SIGNAL	2.5	2.5	0
	B7	NOT USED			
DK GRN	B8	A/C SIGNAL	0	0	0
TAN	B9	P.S.P.S.	12	12	0
ORN/BLK	B10	PARK/NEUTRAL SW. SIGNAL	0	0	0
	B11	NOT USED			
	B12	NOT USED			

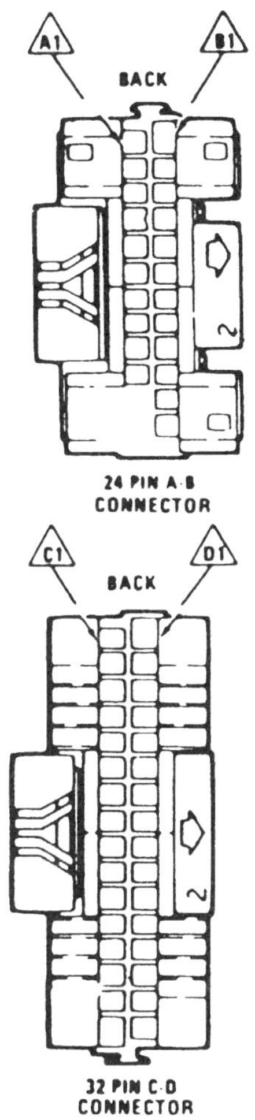

IGN. "ON"	ENG RUN	OPEN CKT.	CIRCUIT	PIN	WIRE COLOR
			NOT USED	C1	
			NOT USED	C2	
			IAC B LO	C3	LT GRN/ BLK
NOT USABLE			IAC B HI	C4	LT GRN/ WHT
			IAC A HI	C5	LT BLU/ WHT
			IAC A LO	C6	LT BLU/ BLK
			NOT USED	C7	
0	0	0	4TH GEAR SIGNAL	C8	DK GRN
0	0	0	A/C FAN REQUEST	C9	DK GRN
1.9	1.7	0	COOLANT TEMP SIGNAL	C10	YEL
			NOT USED	C11	
.55 V ± .05 V			TPS BACK-UP	C12	DK BLU
.55 V ± .05 V			TPS SIGNAL	C13	DK BLU
5	5	0	TPS 5V REF	C14	GRY
12	14	12	B INJ 2 4 6	C15	LT GRN
12	14	12	BATT 12 VOLTS	C16	ORN

WIRE COLOR	PIN	CIRCUIT	IGN. "ON"	ENG. RUN	OPEN CRT.
BLK	D1	GRN'D			
DK GRN/ WHT	D2	COOLING FAN CONTROL	12	14	12
	D3	NOT USED			
	D4	NOT USED			
TAN/BLK	D5	EST BYPASS	0	4.75	0
TAN	D6	GRN'D (O_2)	0	0	0
PPL	D7	O_2 SENSOR SIGNAL			
DK GRN WHT	D8	EGR DIAG. SWITCH	12	14	0
	D9	NOT USED			
	D10	NOT USED			
	D11	NOT USED			
	D12	NOT USED			
BLK	D13	COOLANT, TPS, MAT, SENSOR GRD	0	0	0
LT GRN	D14	INJ 2 4 6 "B"	12	14	12
LT BLU	D15	INJ 1 3 5 "A"	12	14	12
LT BLU	D16	INJ 1 3 5 "A"	12	14	12

ENGINE 2.8L LB6 TYPICAL

Figure 1-14. Pin voltage charts provide information for checking signal voltages at the PCM harness connector.

Troubleshooting Intermittent Problems

Intermittent driveability problems can often be extremely difficult to diagnose and repair. If you are lucky, the intermittent problem will set a soft code in PCM memory. This gives you a clue, at least, about the general area in which to start testing. Remember, however, that if you clear the codes, the problem may not recur right away. You may have to simulate the conditions that cause the problem or road test the vehicle to catch the intermittent fault. The following paragraphs outline some basic points that can help you troubleshoot intermittent problems.

Try To Simulate The Problem

Try to recreate the conditions that the customer describes. You may not always be able to duplicate the conditions exactly, but try to come as close as possible. You may have to road test the vehicle, or you may have to leave it standing overnight to recreate a cold-starting problem. If the problem caused a soft code, try to get the code to recur during testing.

Wiggle Tests and Output Cycling Tests

Most control systems have long-term memory that will record soft codes for intermittent problems. In a wiggle test, you tap or wiggle wiring, connectors, and components to try to get the problem to occur and reset the code. Wiggle tests can be done with the key on and the engine off or with the engine running. You can usually conduct a wiggle test with a scan tool connected, this allows you to monitor the data stream and watch for parameters to suddenly change. Be aware, some systems will not set codes with scan tool connected and operating in the diagnostic mode. For these vehicles, you can either monitor the data stream, or simply perform the test without the scan tool. If you do get the problem to occur, remember what you did to cause it, or when the code set. After wiggle testing, check for codes in memory once again to verify the fault.

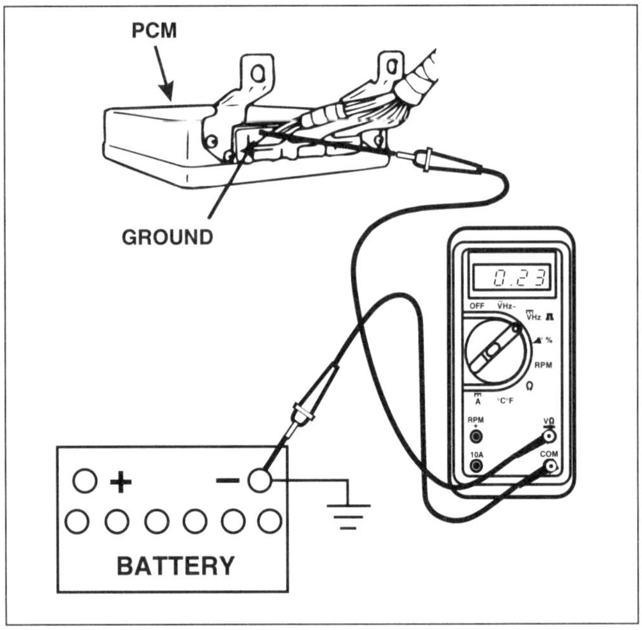

Figure 1-15. Check for a voltage drop at the PCM ground connection with a DVOM.

Some vehicles have output cycling tests as part of their diagnostic capabilities. These systems have a bidirectional PCM that you can command with a scan tool. When in the test mode, the PCM generates a signal to switch an actuator on and off in a controlled condition. While the actuator is being switched, you can perform a wiggle test and take voltmeter readings as you try to pinpoint an intermittent fault.

Check Connectors For Damage

Intermittent problems are often caused by damaged wiring connectors and terminals. Poor connections create high resistance that interrupts the signal. Unplug the connectors in the problem circuit and inspect them carefully. Look for:

- Bent or broken terminals
- Corrosion
- Terminals that have been forced back into the connector shell, causing an intermittent connection
- Loose, frayed, or broken wires in the connector shell

Most connectors on vehicle control systems are repairable, and often spraying a small amount of electronic contact cleaner on the terminals will help to remove dirt, corrosion, and moisture. If damage is severe, replace the connector.

Road Test And Record Data

If the vehicle transmits computer data in a road test or normal operating condition, drive it and try to duplicate the problem. A quality scan tool will allow you to electronically record the data stream when the problem occurs. When you return to the shop, analyzing the recorded data can help locate the faulty circuit or component. Look for unusual circuit activity, such as high or low signals, voltage dropout, and lack of switching. When you spot a problem on one circuit, be sure to check for unusual activity on all of the related circuits.

1. A car is towed to the shop because the engine will not crank.

   Technician A says the problem could be due to an open in the clutch switch.

   Technician B says the problem may be due to the PCM not receiving a pulse from the crank angle sensor.

   Who is right?

   a. A only
   b. B only
   c. Both A and B
   d. Neither A nor B

   Explanation: An open in the clutch switch will prevent a car from cranking, but a faulty crank angle sensor will not. A faulty crank angle sensor will allow the engine to crank but it will not allow it to start, so a is the correct answer.

2. The composite vehicle just failed an I/M 240 test due to high emissions. The O2 sensor voltage readings are higher than normal and no codes are set.

   Technician A says this could be caused by an open circuit at one of the PCM injector drivers.

   Technician B says this could be caused due to a ruptured fuel pressure regulator diaphragm.

   Who is right?

   a. A only
   b. B only
   c. Both A and B
   d. Neither A nor B

   Explanation: A ruptured fuel pressure regulator diaphragm will cause the fuel pressure to increase. As a result, the vehicle will run rich and the O2 sensor voltage will be high. Answer a is wrong because an open circuit at the injector driver will prevent the injector from turning on, thus causing a lean fuel condition and low O2 sensor voltage. Technician B is correct.

3. A compression test has been made on an inline 6-cylinder engine. Cylinders 3 and 4 have readings of 10 psi. All other cylinders have readings between 130 and 135 psi.

   Technician A says this could be caused by a blown head gasket.

   Technician B says it could be caused by wrong valve timing.

   Who is right?

   a. A only
   b. B only
   c. Both A and B
   d. Neither A nor B

   Explanation: A blown head gasket between cylinders 3 and 4 could easily cause low readings on those two cylinders, but wrong valve timing would have an effect on all the cylinders. Technician A is correct.

4. A tow truck dropped a no-start vehicle at your shop door. It is a late-model vehicle with a computerized engine management system.

   Technician A says to begin the diagnosis by first retrieving any diagnostic trouble codes with a scan tool.

   Technician B says it is more efficient to begin the diagnosis by first checking for secondary spark.

   Who is right?

   a. A only
   b. B only
   c. Both A and B
   d. Neither A nor B

   Explanation: The powertrain control module (PCM) monitors many automotive circuits, and stores diagnostic trouble codes (DTC) for those failing to meet preset parameters. Since the PCM does not monitor many mechanical and ignition related operations, the technician should begin by checking for secondary spark. If secondary spark is present, then the technician should run a compression or fuel pressure check. Technician B is correct.

5. A customer brings his overheating car into your shop. The customer tells you that the car overheats while driving only on the freeway.

   Technician A says that the lower radiator hose may be collapsing.

   Technician B says that the water pump blades may be loose on the water pump shaft.

   Who is right?

   a. A only

   b. B only

   c. Both A and B

   d. Neither A nor B

   Explanation: The demands on the coolant system increase as engine speed increases. A faster rotating water pump shaft will create lower pressures at the pump's inlet. To prevent the collapse of the inlet side hose, manufacturers usually install a supporting spring. Collapse of the hose due to a corroded spring usually only affects cruise operation. Loose water pump blades would increase coolant temperature, regardless of engine speed. Technician A is correct.

6. A customer comes in to your shop complaining that his late- model vehicle has failed the I/M 240 emissions test with the following emission readings:

HC g/m	1.06	
CO g/m	25.50	
NO_x	ppm	0.752
CO_2	ppm	395

   Technician A says the aftermarket rear differential ratio of over 4.11 to 1 could be contributing to the cause of higher emissions during the dynamic portion of the test.

   Technician B says a faulty vehicle speed sensor is not allowing the torque converter clutch to operate, causing the emissions to be higher than the FTP (Federal Test Procedure) level during the drive cycle.

   Who is right?

   a. A only

   b. B only

   c. Both A and B

   d. Neither A nor B

   Explanation: Any radical change in the differential gear ratio, either up or down, can cause an increase in the emissions. Also, if the torque converter clutch fails to engage at highway speeds, it will cause a slight emissions increase, so c is the answer.

Chapter Two

COMPUTERIZED ENGINE CONTROLS DIAGNOSIS

This chapter covers powertrain control module (PCM) circuits, input and output devices, and their interrelationships. It includes most of the common PCM circuits as well as all those used in the ASE composite vehicle. It describes their operation and explains the driveability problems and emission failures that a malfunction of each may cause.

DRIVEABILITY PROBLEM AND EMISSION FAILURE DIAGNOSIS

For each of the following sensors and actuators we briefly define or explain its use, list typical serial data (if any is normally available), and describe the common driveability or emissions problem a failure of this part would cause.

Sensors

Sensors monitor engine parameters, operating conditions, and engine component functions. Sensors are also input devices that transmit signals to the PCM (Powertrain Control Module) describing the state, temperature, position, or operating status of the system or part they monitor.

Engine Speed Sensors

Engine speed (RPM), crankshaft position, and camshaft position can be transmitted to a PCM in the following ways:

- A pickup coil (permanent magnet) sensor.
- A magnetic hall-effect switch
- An optical sensor

Pickup coil

Pickup coil sensors are also called variable reluctance sensors or permanent magnet generators. They are voltage generating sensors, and work like pickup coils used in electronic distributors. The pickup coil is wound around a permanent magnet, figure 2-1. As the teeth of a rotating trigger wheel pass by the magnet, the magnetic field expands and collapses to generate an alternating current (AC) voltage. As the speed of the engine increases, so does the frequency and am-

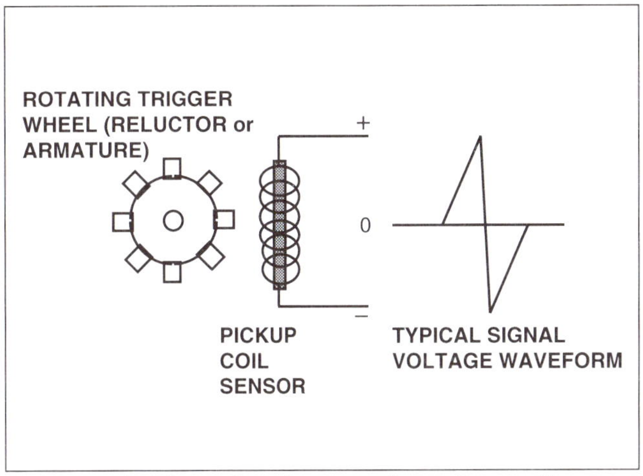

Figure 2-1. A pickup coil produces waveform voltage as the magnetic field expands and collapses.

plitude (voltage output) of the signal. You can check the signal from this type of sensor with a digital volt-ohm meter (DVOM) on the AC volt or frequency scale. You can also check the resistance of the pickup coil winding, using the ohm scale. However, checking pickup coils with a lab scope is much more accurate for picking up any glitches in the signal.

Hall-effect

A hall-effect switch generates a digital signal (on/off), figure 2-2. A hall-effect sensor usually has three wires, unless it is a combination-type sensor. The hall element receives an input voltage to power up the element. The PCM supplies a signal voltage, usually 5 or 12 volts that is switched on and off ground as the shutter blade rotates through the hall sensor. When a shutter vane is between the element and the magnet, the magnetic field is blocked by the shutter blade and the signal voltage will be high. When the shutter blade is not between the element, the magnetic field is generated across the element causing the signal voltage to go low. The third wire provides a ground path.

Optical sensors

Optical sensors also generate a digital signal (on/off), but operate differently than a hall-effect switch. An op-

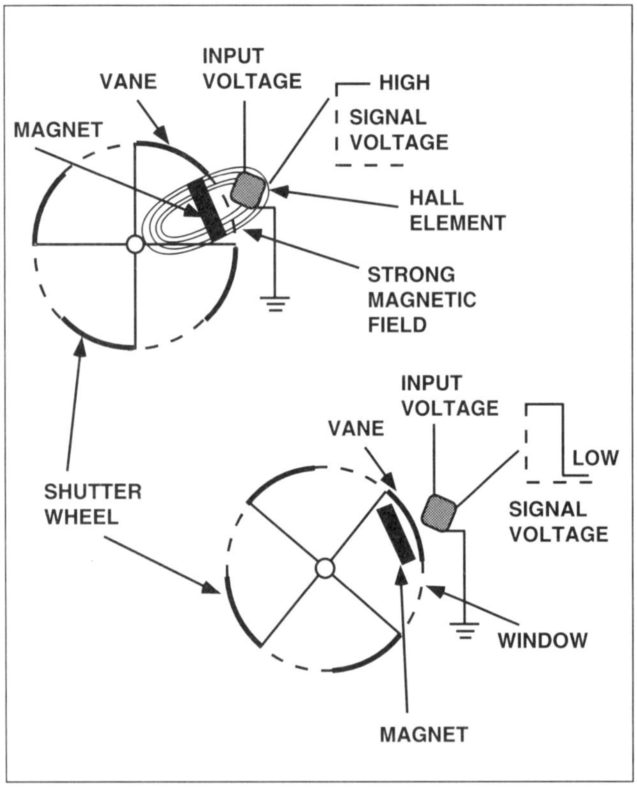

Figure 2-2. A Hall-Effect switch produces a high signal when the magnetic field is strong, and a low signal when the field is weak.

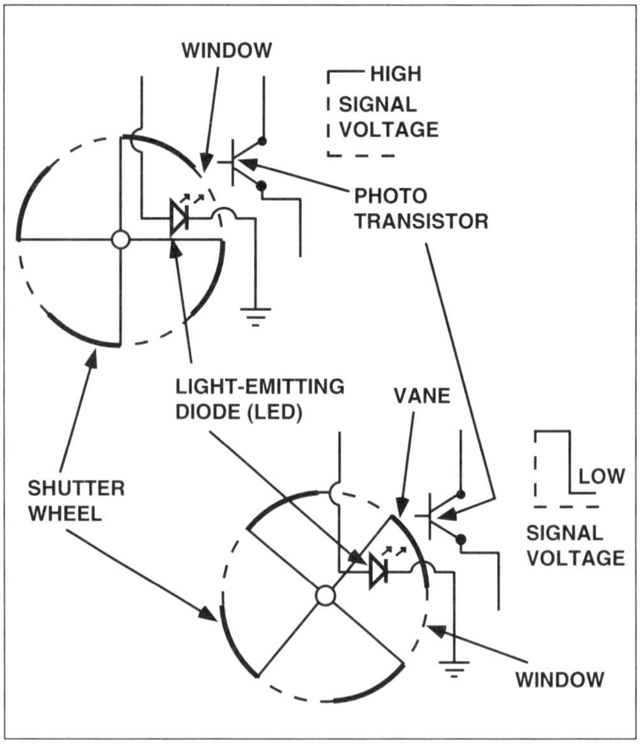

Figure 2-3. An optical sensor uses an LED, phototransistor, and shutter wheel to produce a digital signal.

tical sensor uses a light-emitting diode (LED) and a phototransistor to toggle a signal voltage supplied by the PCM on and off, figure 2-3. An optical sensor requires an input power voltage and a PCM-generated signal voltage in order to produce an accurate signal. An optical sensor usually contains three wires, unless it is a combination-type sensor with a high and low data rate signal.

The PCM supplies a signal voltage, usually 5 volts, which switches on and off as the rotating wheel or blade passes between the phototransistor and LED, toggling the ground circuit. When the rotating reflector directs light to the phototransistor, the light turns off the base of the transistor and causes the signal voltage to stay high. When the reflector plate blocks the light to the phototransistor, the base of the transistor is on and signal voltage stays low. The third wire is a ground path.

Engine speed sensors can be driven off the distributor or the crankshaft. If the speed sensor is crank driven, information supplied to the PCM can only be RPM and companion cylinders. To detect #1 cylinder for sequential fuel injection, a position signal must be supplied to either the distributor or an additional camshaft sensor. Engine sensor data may show up in a data stream as: Engine RPM.

Speed sensor driveability problems and emission failures

Engine speed sensors are the only signal that cannot be backed up by a substitute signal. If you loose the engine speed signal, the injectors will not trigger and the vehicle will not start. Diagnostic Trouble Codes (DTC's) may or may not set for these types of failures. The following are examples of some of the driveability problems and emission failures that can occur with speed sensor circuit problems.

- No toggling of voltage on PCM will cause a no-start
- Erratic voltage signals can cause rough idle, engine stalling, misfire, cutout, hesitation, excessive hydrocarbons, and possible no-start.

Manifold Absolute Pressure Sensor

The manifold absolute pressure (MAP) sensor is a piezoresistive crystal that has a sealed reference on one side, figure 2-4. The other side is connected to intake manifold vacuum. MAP sensors send an analog voltage signal that is proportional to air pressure (or vacuum). The signal voltage is low, approximately 0.5 volts, at idle (high vacuum) and signal voltage is high, about 4.75 volts, with no vacuum (key on, but the engine off).

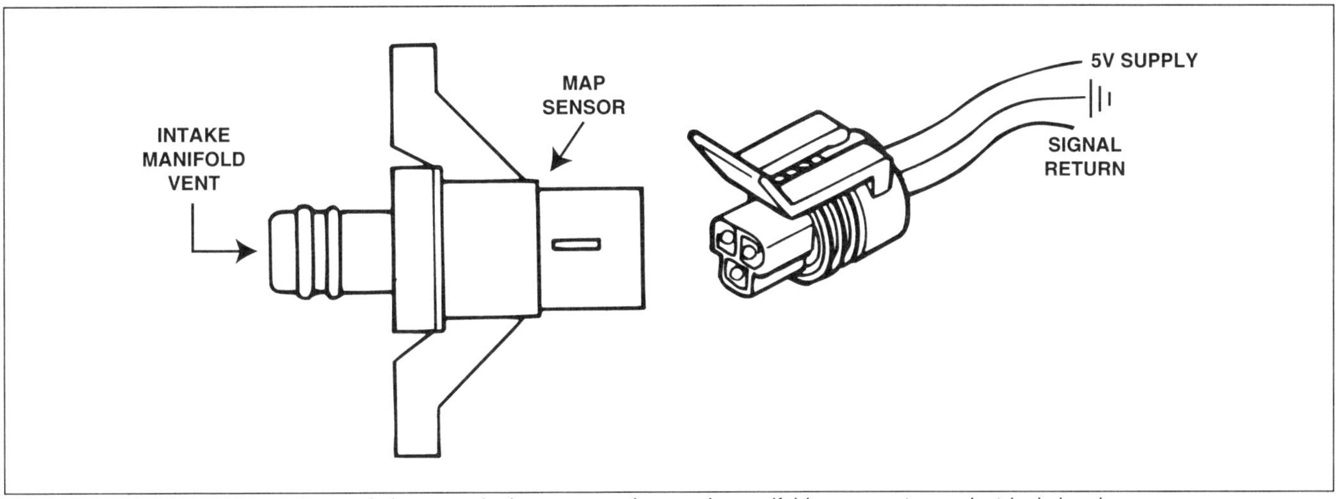

Figure 2-4. MAP sensors are piezoresistive crystals that convert changes in manifold pressure to an electrical signal.

Some MAP sensors work under the same principal but produce a frequency instead of an analog voltage. In this case, frequency is high with (low vacuum) and low at engine idle (high vacuum).

The PCM uses MAP sensor input primarily for fuel delivery and ignition timing control. As engine vacuum decreases under engine load, the PCM receives a higher voltage input. The PCM increases injector pulse width proportionally, based on the high input voltage. Increased MAP voltage also affects ignition timing. As MAP voltage decreases, the PCM may advance the ignition timing. Timing is based on internal PCM calibration tables or look up tables.

The PCM also monitors the MAP sensor for rate (speed) of voltage change. The MAP sensor usually has equal authority to that of the TP sensor on rapid acceleration. A MAP sensor will also back up the TP sensor should the circuit fail, placing the vehicle into a back-up or limp-in mode. MAP sensor data may show up in a data stream in several different ways:

- Digital Voltage
- Inches of mercury. Either vacuum or manifold absolute pressure
- Kilo Pascals (kPa). Manifold absolute pressure.

MAP driveability problems and emission failures

The following are examples of some of the driveability problems and emission failures that can occur with MAP sensor circuit problems.

- Improper voltage reading indicating a low manifold vacuum can cause a rich mixture, erratic idle, engine stalling, high emissions, and retarded ignition timing.
- Improper voltage reading indicating a high manifold vacuum can cause a lean mixture, erratic idle, hesitation, high emissions, and advanced ignition timing.
- A signal that glitches high (towards reference voltage) at idle can cause a rich mixture, an erratic idle, engine stalling, and possible no-start.
- A signal that glitches to a voltage lower than minimum specifications at idle can cause a lean mixture, an erratic idle, and engine stalling.
- A signal that glitches high (towards reference) voltage at cruise can cause a rich mixture, hesitation, and surge.
- A signal that glitches to a lower voltage than minimum specifications at cruise can cause a lean mixture, hesitation, surge, and stumble, and detonation.

Remember, these conditions can be caused by both electrical problems within the circuit or by mechanical problems. Relevant mechanical problems include: low manifold vacuum, a plugged catalytic converter, vacuum leaks, or an open or a restricted vacuum line.

Throttle Position Sensor

The throttle position (TP) sensor is a potentiometer that sends an analog variable voltage signal proportional to the throttle opening, figure 2-5. TP sensors monitor both motion and position. Generally TP sensors operate on a 5 volt reference. The signal voltage is low at closed throttle, approximately 0.5 to 1.1 volts. As the throttle opens, voltage gradually increases to approximately 4.75 volts at wide open throttle.

The PCM uses the TP sensor input primarily for fuel delivery and ignition timing control. As the throttle opening increases, the PCM receives a higher voltage input. The PCM then increases the injector pulse width accordingly, based on the input voltage. Increased TP sensor voltage also affects ignition timing. As TP

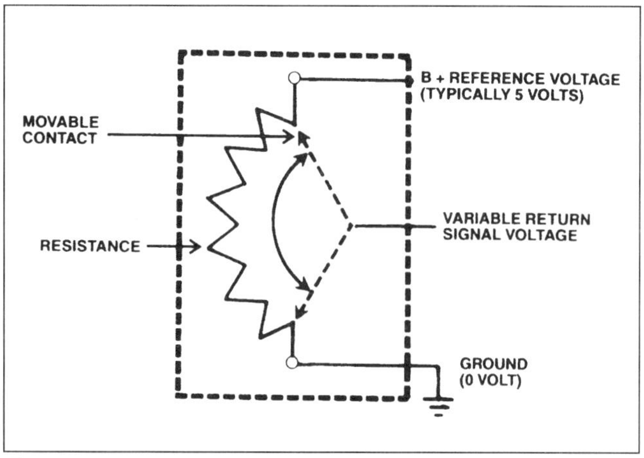

Figure 2-5. A potentiometer uses a movable contact to vary resistance and send an analog signal.

sensor voltage increases, the PCM may advance the ignition timing based on PCM calibration tables. The PCM monitors the TP sensor for rate (speed) of voltage change, as well as normal throttle movement. The TP sensor will usually have equal authority to that of the MAP or MAF sensor on rapid acceleration. A TP sensor will also back up MAF or MAP sensors should one of those circuits fail, placing the vehicle into a back-up or limp-in mode. TP sensor data may show up in a data stream in several different ways:

- Digital Voltage
- Percentage (%) of Throttle Position
- Indicator C/T (Closed Throttle), P/T (Part Throttle), W/T (Wide Open Throttle)
- Minimum TP sensor voltage

TP sensor driveability problems and emission failures

The following are examples of some of the driveability problems and emission failures that can occur with TP sensor circuit problems.

- Improper adjustment can cause an erratic idle, engine stalling, off-idle hesitation, and high emissions.
- Improper adjustment can also cause an automatic transmission to shift erratically.
- High voltage may prevent starting because it places the system in "clear flood" mode.
- A signal that glitches high (towards reference voltage) at idle can cause a rich mixture, an erratic idle, and engine stalling.
- A signal that glitches to a voltage lower than minimum specifications at idle can cause a lean mixture, an erratic idle, and engine stalling.
- A signal that glitches high (towards reference) voltage at cruise can cause a rich mixture, hesitation, surge, and detonation.
- A signal that glitches to a lower voltage than minimum specifications at cruise can cause a lean mixture, hesitation, surge, and stumble.
- A signal glitch, either high or low, at cruise may also cause erratic torque converter clutch operation.

Engine Coolant Temperature Sensor

The engine coolant temperature (ECT) sensor is a thermistor type solid state variable resistor that usually has two connector terminals. Most ECT sensors are negative temperature coefficient (NTC) type, which decrease resistance as temperatures increases. The PCM applies a reference voltage (usually 5 volts) to the thermistor. In some cases, a pull-up resistor inside the PCM applies the reference voltage, figure 2-6.

The PCM monitors the voltage drop across the thermistor and interprets any change in voltage as a signal of changing temperature. When the engine temperature is cold, the sensor drops most of the reference voltage across the thermistor. The signal voltage measured ahead of the thermistor by the PCM is high. When engine temperature is hot, NTC thermistor voltage is low because it drops only a small amount of reference voltage and the signal voltage measured ahead of the thermistor is low. Regular thermistor type circuits (non pull-up type) will operate at about 0.25 volts at 248°F (120°C) to 4.90 volts at -40°F (-40°C).

The PCM uses ECT input during cold engine operation for idle speed control, fuel enrichment, and ignition timing advance. As the engine reaches operating temperature, the ECT will influence closed loop operation. During normal engine temperature, the ECT influence is diminished by inputs from TP sensor, MAP, or MAF. If a failure occurs in the ECT circuit, the intake air tem-

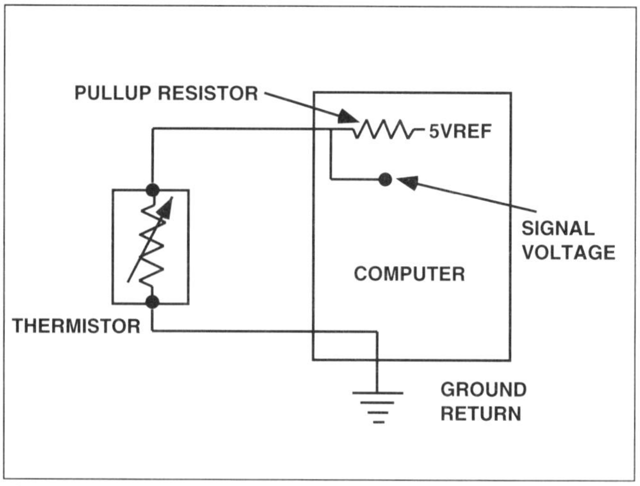

Figure 2-6. The computer applies a 5-volt reference to a thermistor through a pullup resistor.

perature sensor circuit will back the ECT up. ECT sensor data may show up in a data stream in several different ways:

- Digital Voltage
- Degrees Fahrenheit
- Degrees Celsius

ECT driveability problems and emission failures

The following are examples of some of the driveability problems and emission failures that can occur with ECT sensor circuit problems.

- A false cold temperature signal on a warm engine could cause a high idle, excessive injector pulse width, over advanced ignition timing, and high emissions
- A false warm temperature on a cold engine could cause low idle speed, a lean mixture, a lack of power due to inadequate ignition timing advance, and high emissions.
- A signal that glitches to a high voltage at cruise can cause a rich mixture, hesitation, surge, and stumble.
- A signal that glitches to a low voltage at cruise can cause a lean mixture, hesitation, surge, and stumble.
- a false high temperature signal at idle signaling a overheat condition can cause the electric cooling fan to run continuously.

Intake Air Temperature Sensor

The intake air temperature (IAT) sensor is a thermistor type circuit. The operation of this circuit is identical to that of the ECT circuit. The negative temperature coefficient thermistor (NTC) decreases resistance as temperatures increases, figure 2-7. The PCM applies a reference voltage (usually 5 volts) to the thermistor. In some cases the 5 voltage reference may be applied through a pull-up resistor inside the PCM. Regular thermistor type circuits (non pull-up type) will operate at about 0.25 volts and 248°F (120°C) to 4.90 volts at -40°F (-40°C).

The PCM uses IAT input during cold engine operation for idle speed control and fuel enrichment. As the engine reaches operating temperature, IAT influence is diminished by inputs from the ETC, TP, MAP, or MAF Sensors. If a failure occurs in the IAT circuit, the ETC circuit will back the IAT up. IAT sensor data may show up in a data stream in several different ways:

- Digital Voltage
- Degrees Fahrenheit
- Degrees Celsius

	TEMPERATURE DEGREES F(C)	SENSOR RESISTANCE OHMS	SENSOR VOLTAGE DROP
	−40(−40)	100,000+	5.00
	18(−8)	3260	3.93
	32(0)	2940	3.56
	50(10)	2445	2.98
	68(20)	1956	2.41
	86(30)	1493	1.86
SWITCH TO SHUNT ←	104(40)	1115	1.40
	122(50)	786	3.69
	140(60)	566	3.27
	158(70)	426	2.87
	176(80)	308	2.44
	194(90)	226	2.05
	212(100)	170	1.70
	230(110)	128	1.39
	248(120)	98	1.15

Figure 2-7. Operating range chart for a dual-range thermistor with a shunt circuit.

IAT driveability problems and emission failures

The following are examples of some of the driveability problems and emission failures that can occur with IAT sensor circuit problems.

- A false cold temperature signal on a warm engine could cause a high idle, excessive injector pulse width, and high emissions.
- A false warm temperature on a cold engine could cause low idle speed, lean condition, and high emissions.
- A signal that glitches to a high voltage at cruise can cause a rich mixture, hesitation, surge, and stumble.
- A signal that glitches to a low voltage at cruise can cause a lean mixture, hesitation, surge, and stumble.
- a false high temperature signal at idle signaling a overheat condition can cause the electric cooling fan to run continuously.

Oxygen Sensor

The oxygen sensor (O2S) is a galvanic battery, and most are made of zirconium dioxide (zirconia). An O2S is capable of generating an output voltage of approximately 1 volt. The signal voltage range is usually from 0.1 to 0.9 volt (100 to 900 mv). The difference between the amount of oxygen in the exhaust compared to the amount of oxygen in the outside air generates the voltage, figure 2-8.

A simple procedure for O2S diagnosis is to remember the phrase "lean and low". When the exhaust oxygen content is high, there is somewhat less difference between the oxygen in the air and the oxygen in the exhaust. Therefore, the sensor will produce a lower volt-

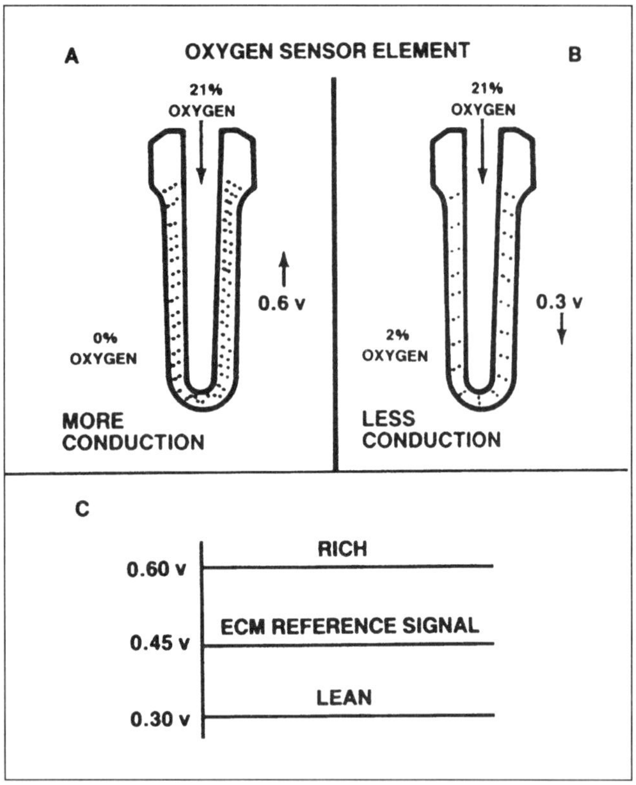

Figure 2-8. A difference in oxygen content between the atmosphere and exhaust will generate voltage from an oxygen sensor.

age (100mv to 420 mv), which indicates a lean condition to the PCM.

When exhaust oxygen content is low there is a large difference between the oxygen in the air and the oxygen in the exhaust. Therefore, the sensor will produce a higher voltage (460mv to 900 mv), which indicates a rich condition to the PCM.

The sensor continually switches between high and low voltage, signaling the PCM to maintain the ideal air/fuel ratio. An O2S voltage of 440 to 550 mv is equivalent to about a 14.7 to 1 air/fuel ratio. An O2S must be at 572°F (300°C) before it generates a valid signal.

To generate this voltage signal more quickly and accurately, most vehicles use Heated Oxygen Sensors (HO2S). Heated oxygen sensors use positive coefficient heaters which build resistance as heat is generated. A HO2S can heat to about 1000°F (538°C) in as little as eight seconds. This allows the carmaker to locate an HO2S further down stream in the exhaust for more accurate readings of exhaust oxygen content. The exhaust temperature needed to keep the sensor at operating temperature is now not as critical.

The PCM uses HO2S or O2S input during closed loop operation for air/fuel ratio control. To best diagnose this circuit, observe O2S cross counts on a DVOM with an analog bar or an oscilloscope. Watch the sensor cross over the 450 mv range. HO2S and O2S sensor data may show up in a data stream in several different ways:

- Digital Voltage
- Cross Counts
- Lean/Rich (LED lights)
- Exhaust Lean/Center/Rich (LCD Display)
- Open/closed loop (other inputs also effect this reading)

O2S driveability problems and emission failures

The following are examples of some of the driveability problems and emission failures that can occur with oxygen sensor and circuit problems.

- Sensor voltage constantly low (lean) can cause the injectors to increase the injector pulse width, causing a rich mixture, poor idle, hesitation, surging, high emissions, and catalytic converter damage.
- Sensor voltage constantly high (rich) can cause the PCM to narrow the injector pulse width, causing a lean mixture, poor idle, hesitation, surging, high emissions, and catalytic converter damage.
- Lazy sensor voltage can cause incorrect injector pulse width, poor idle, surging, high emissions, and catalytic converter damage.
- A faulty O2 heater or circuit causes an O2 sensor to cool down at idle, making the injectors increase injector pulse width. The result is a rich mixture, poor idle, hesitation, high emissions, and possible catalytic converter damage.

Speed Sensor

The vehicle speed sensor (VSS) transmits a signal in proportion to vehicle speed, figure 2-9. Any of the following will produce a speed signal.

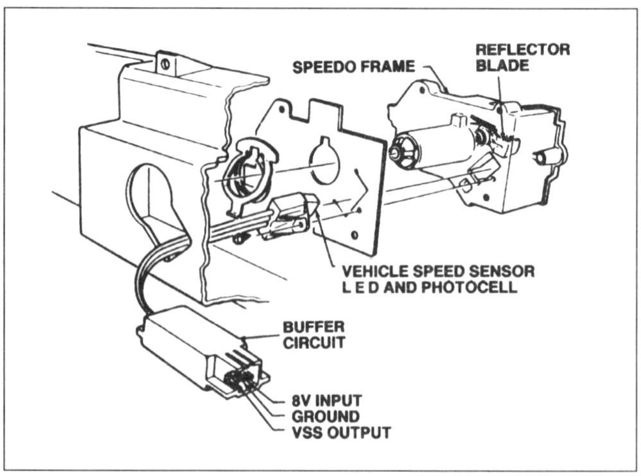

Figure 2-9. An LED and a photocell in the speedometer act as a speed sensor in this application.

- A pickup coil (permanent magnet) sensor.
- An optical sensor in the speedometer head.
- A magnetic reed switch.

The VSS generates a signal that increases in frequency as vehicle speed increases. The PCM uses VSS input to control torque converter clutch operation and high speed fuel cutoff. A VSS signal may also be used as an input for electronic cruise control systems. VSS data may show up in a data stream in several different ways:

- Miles Per Hour (MPH)
- Kilometer Per Hour (KPH)

VSS driveability problems and emissions failures

The following are examples of some of the driveability problems and emission failures that can occur with VSS circuit problems.

- Improper high frequency reading can cause premature torque converter lockup. This can be caused by a mechanical fault, electronic fault, or something as simple as the wrong tire size on the vehicle.
- Improper low frequency reading can delay torque converter lockup due to a false vehicle speed signal. This can be caused by a mechanical fault, electronic fault, or the wrong tires size on the vehicle. Delayed or prolonged torque converter lockup will also reduce fuel economy and increase exhaust emissions.
- An erratic signal can cause torque converter lock up problems.

Barometric Pressure Sensor

The barometric pressure (BARO) sensor is a piezoresistive crystal that senses atmospheric pressure, figure 2-10. BARO sensors send an analog voltage signal that is proportional to atmospheric pressure. The signal voltage is high at sea level, approximately 3.8 to 4.8 volts and low at high elevations, about 2.5 to 3.5 volts at 9000 feet (2700 meters). These readings are true with the key on, whether the engine is running or not.

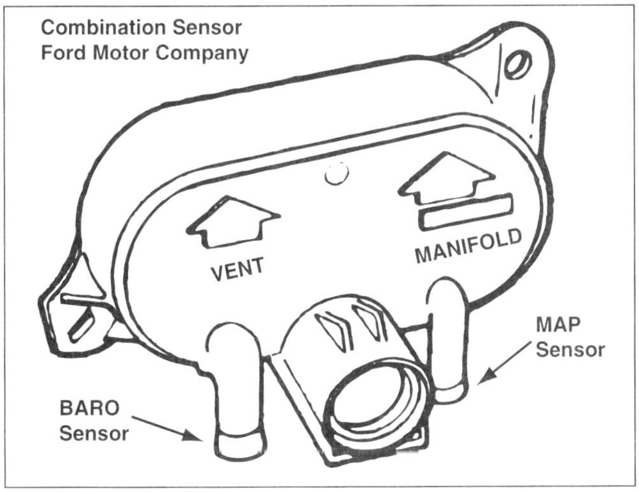

Figure 2-10. BARO sensors are piezoresistive crystals that convert changes in atmospheric pressure to an electrical signal.

Some BARO sensors work under the same principal but produce a frequency instead of an analog voltage. In this case, the frequency is high at low altitude and low at high altitude or high vacuum conditions, such as idle speed.

The PCM uses BARO sensor input primarily for fuel delivery and ignition timing control. As atmospheric pressure changes, the BARO produces a voltage change, which is sensed by the PCM. The PCM then increases or decreases injector pulse width accordingly, based on the input voltage. Changes in BARO voltage will also effect ignition timing. BARO sensor data may show up in a data stream in several different ways:

- Digital Voltage
- Inches of mercury. Either vacuum or manifold absolute pressure
- Kilo Pascals (kPa). Manifold absolute pressure.

BARO driveability problems and emission failures

The following are examples of some of the driveability problems and emission failures that can occur with BARO sensor circuit problems.

- Improper voltage reading can cause an erratic idle, engine stalling, off-idle hesitation, loss of power going up in elevation, and high emissions.
- A signal that glitches high (towards reference voltage) at idle can cause a rich mixture, an erratic idle, engine stalling, and possible no-start.
- A signal that glitches to a voltage lower than minimum specifications at idle can cause a lean mixture, an erratic idle and engine stalling.
- A signal that glitches high (towards reference) voltage at cruise can cause a rich mixture, hesitation, and surge.
- A signal that glitches to a lower voltage than minimum specifications at cruise can cause a lean mixture, hesitation, surge, and stumble.

Remember, these problems can be caused by both electrical problems within the circuit or by a mechanical failure, such as a plugged sensor vent port.

Mass Airflow Sensor

The mass airflow (MAF) sensor is a heated wire or a heated film sensor. MAF sensors transmit an analog voltage signal or frequency signal. The signal is proportional to the weight (molecular mass) of the air entering the engine. Voltage or frequency increases as more air is drawn into the engine during throttle opening, figure 2-11.

The PCM uses MAF sensor input primarily for fuel delivery and ignition timing control. As more air is drawn into the engine under load, the PCM receives a higher voltage or frequency (high signal) input. The PCM then increases injector pulse width accordingly, based on the higher signal. The increased MAF signal also affects ignition timing. As the MAF signal increases, the PCM may advance the ignition timing based on PCM calibration tables. The MAF sensor usually has equal authority to that of the TP sensor on rapid acceleration. A MAF sensor will also back up the TP sensor should the circuit fail, placing the vehicle into a back-up or limp-in mode. MAF sensor data may show up in a data stream in several different ways:

- Digital Voltage
- Mass Airflow Grams /sec
- Load LV8

MAF driveability problems and emission failures

The following are examples of some of the driveability problems and emission failures that can occur with MAF sensor circuit problems.

- Improper signal reading can cause an erratic idle, engine stalling, off-idle hesitation, high emissions, and possible no-start.
- A signal that glitches high at idle can cause a rich mixture, an erratic idle, engine stalling, and possible no-start.
- A signal that glitches lower than minimum specifications at idle can cause a lean mixture, an erratic idle, and engine stalling.
- A signal that glitches high at cruise can cause a rich mixture, hesitation, and surge.

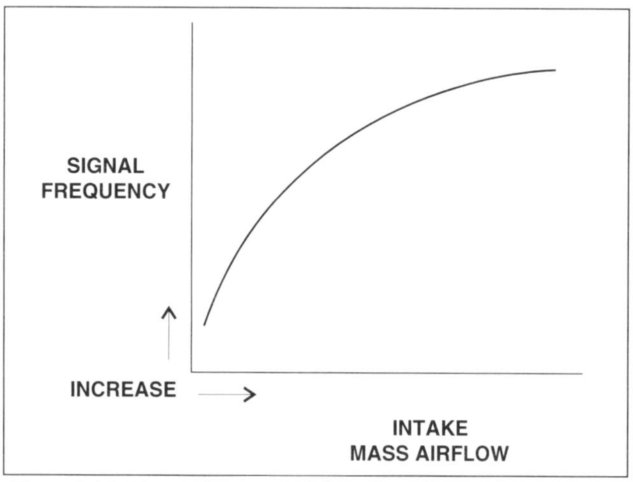

Figure 2-11. The analog mass airflow signal increases in proportion to the mass of the intake air.

- A signal that glitches to a lower than minimum specifications at cruise can cause a lean mixture, hesitation, surge, and stumble.

Remember, either electrical problems within the circuit or mechanical problems can cause these conditions. Relevant mechanical problems include: low manifold vacuum, plugged catalytic converter, intake air leaks behind the MAF sensor, or a restricted intake air passage. Any of these problems can also cause high emissions.

Knock Sensor

The knock sensor (KS) converts the vibrations caused by engine knock into voltage signals which are relayed to the PCM. Most knock sensors are piezoelectric accelerometers tuned to respond to specific the frequencies created by engine knock, figure 2-12. The PCM uses the KS sensor input for ignition timing retard control during engine detonation conditions. The KS may show up in a data stream as:

- Knock on/off
- Knock Retard (degrees)

KS driveability problems and emission failures

A false KS reading at cruise can cause the timing to retard, effecting engine power and performance. This could be caused by other types of drivetrain noise or by a faulty sensor or PCM.

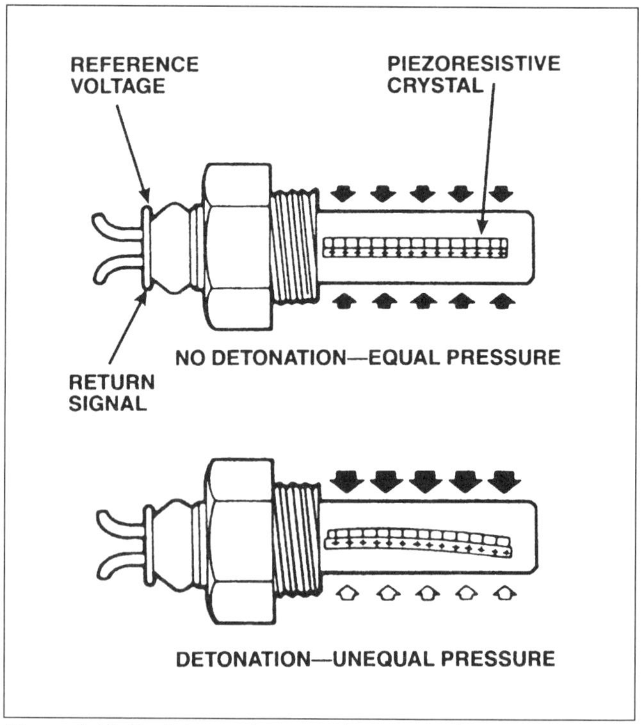

Figure 2-12. Resistance changes when unequal pressure flexes a piezoresistive crystal.

EGR Valve Position Sensor

The EGR valve position (EVP) sensor is a linear, sliding contact potentiometer connected to the top of the EGR valve stem, figure 2-13. The sensor monitors valve stem position, and therefore EGR flow rate. The PCM uses EGR valve position information for acurate control of the EGR vacuum regulated solenoid(s). Usually EVP voltage will be less than 1 volt with the key on and the engine off, to about 4.6 volts at maximum flow. The EVP may show up in a data stream as:

- EVP Voltage
- PFE Voltage

EVP driveability problems and emission failures

The following are examples of some of the driveability problems and emission failures that can occur with EVP circuit problems.

- A EVP voltage that is too low may cause early EGR opening and lead to hesitation, stumble, and engine surging.
- A EVP voltage that is high may cause late or no EGR opening, which can allow engine detonation.
- Intermittent EVP signal may cause hesitation, stumble, surging, and detonation.

Actuators

Actuators are output devices the PCM controls by switching internal driver circuits on and off. The actuators respond to voltage commands which, in turn, control other circuits or convert electrical energy into mechanical work.

Some of the types of actuators used in computerized engine control systems are solenoids, relays, and stepper motors. When diagnosing these output devices, remember that there can be either mechanical or electrical failures of these devices.

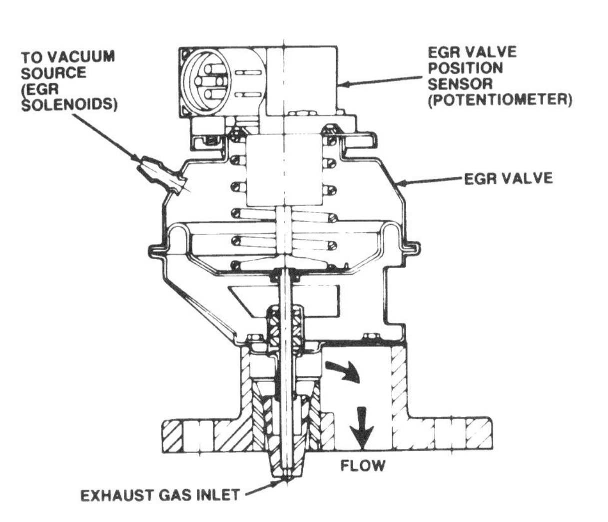

Figure 2-13. Some systems use a potentiometer to sense EGR flow.

Fuel Injectors

Fuel injectors are electrical solenoids mounted in the throttle body or located in the intake manifold, figure 2-14. When the injector receives electrical current, an integral armature and pintle valve move a short distance against a spring to open a small orifice. A constant voltage is applied to the injector and the PCM then grounds the circuit to energize the injector. The amount of time that the PCM turns on the circuit is measured in thousands of a second, or "milli" seconds (ms). The longer the "milli" second duration, the more fuel delivered to the cylinders. This injector on-time is called injector pulse width.

Injectors can be energized all at once, or in groups (group fire injection). Many late-model vehicles use sequential fuel injection (SFI) that energizes the injectors individually in a timed sequence. SFI provides more accurate delivery of fuel, better fuel economy, and lower exhaust emissions. Sequential fuel injection also adds the capabilities of more extensive on-board diagnostics, such as the SFI power balance test and the ability to look at injector pulse width more accurately in a serial data stream. It can also be used for HO2S performance testing on OBD-II compliant vehicles. Figure 2-15 shows a typical injector oscilliscope pattern where:

- "A" represents battery source voltage. It should be steady and have a minimum amount of noise. The injector is off at this point.
- "B" represents the sharp drop to ground caused by the computer completing the circuit. The injector opens at this point. If the drop is slow it indicates a bad driving transistor. If the signal does not reach ground there is high resistance in the computer ground circuit.
- "C" represents the signal rise of 0.5 to 1 volt above ground. This is normal and is caused by resistance in the computer driver circuit. If the voltage rises much above one volt, check the charging system for overcharging and the injectors for low resistance. The rise in voltage marks the transistor opening the circuit and the closing of the injector. This should be a sharp transition; if not, suspect a bad transistor.
- The time between "B" and "C" is the injector pulse width.
- "D" shows the voltage spike, usually about 40 volts, that occurs when the magnetic field of the injector coil collapses. An excessively high spike

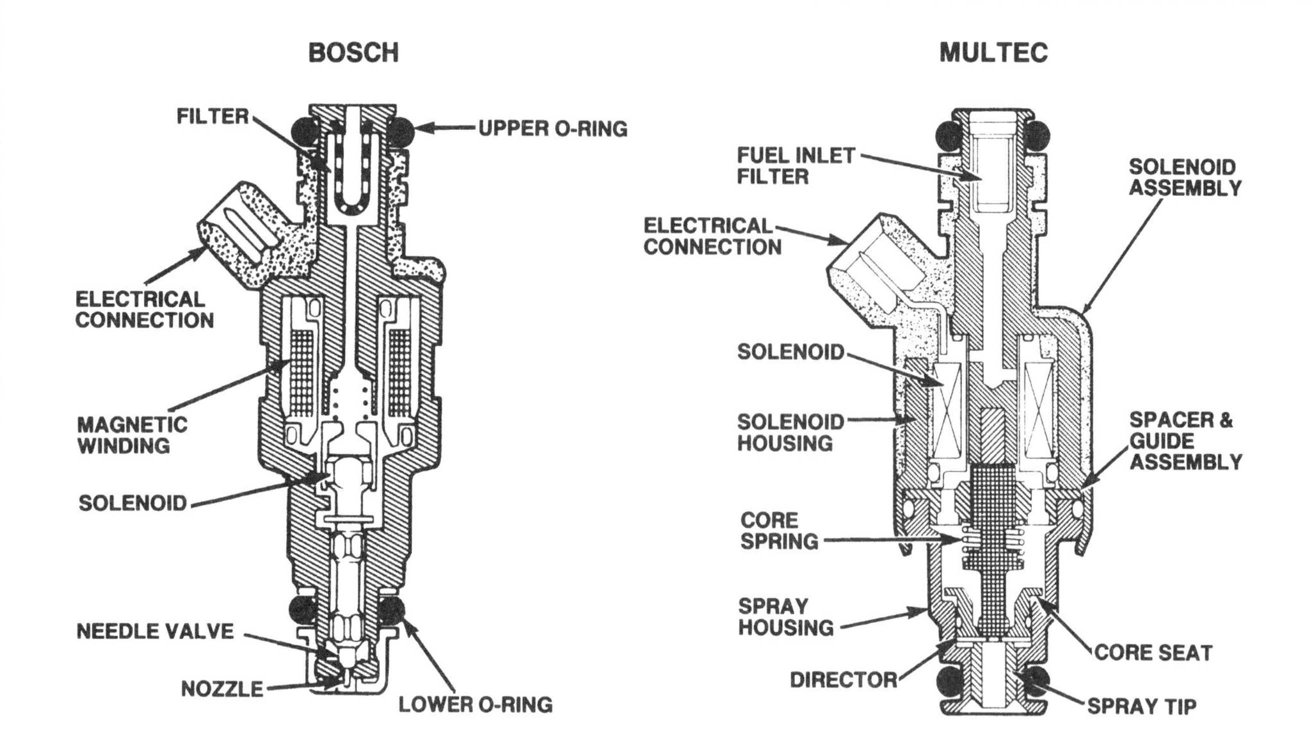

Figure 2-14. Bosch and Rochester Multec fuel injectors.

indicates a charging system overcharge or a bad transistor circuit. A low spike is the result of low system voltage or a bad connections to the injector. If the injector pintle is stuck it will make only a minor difference in the spike, and is usually not detectable.

- "E" represents source voltage.

Fuel management information may show up in a data stream as:

- Injector pulsewidth (ms)
- Block Learn and Integrator (128 + or - 3)
- Long and Short Term Fuel Trim (+ or - a percentage)

Fuel injection circuit driveability and emission failures

The following are examples of some of the driveability problems and emission failures that can occur with injector circuit problems. Problems with the injector parts themselves and other fuel system parts are covered in Chapter 4.

- False inputs to the PCM indicating a lean condition can cause hard starting, poor idle, surging, hesitation, engine detonation, power loss, poor mileage, and high emissions.
- False inputs to the PCM indicating a rich condition can cause, poor idle, surging, hesitation, power loss, poor mileage, and high emissions.

Idle Air Control Valve (IAC)

The idle air control (IAC) valve controls engine idle speeds and prevents stalls due to changes in engine load. The IAC is usually a reversible DC stepper motor that moves in increments, or steps, figure 2-16. The motor moves a shaft back and forth to operate a conical valve. When the conical valve is moved back, more air can bypass the throttle plates and enter the engine, increasing the idle speed. As the conical valve moves inward, the idle speed decreases. Some computerized engine systems use a solenoid operated valve, instead of a stepper motor, to control idle speed.

During idle or deceleration, the PCM calculates the proper position of the IAC valve is calculated by the PCM based on battery voltage, coolant temperature, engine load, vehicle speed, and engine rpm. IAC valve information may show up in a data stream as:

- Idle Air Control Counts (0 to 255)
- Idle Air (a percentage)

IAC driveability problems and emission failures

The following are examples of some of the driveability problems and emission failures that can occur with IAC circuit problems.

- Incorrectly high IAT or ECT (engine cold) signals may cause constant high or an erratic idle.
- Incorrectly low IAT or ECT (engine warm) signals may cause constant hard starting, an errat-

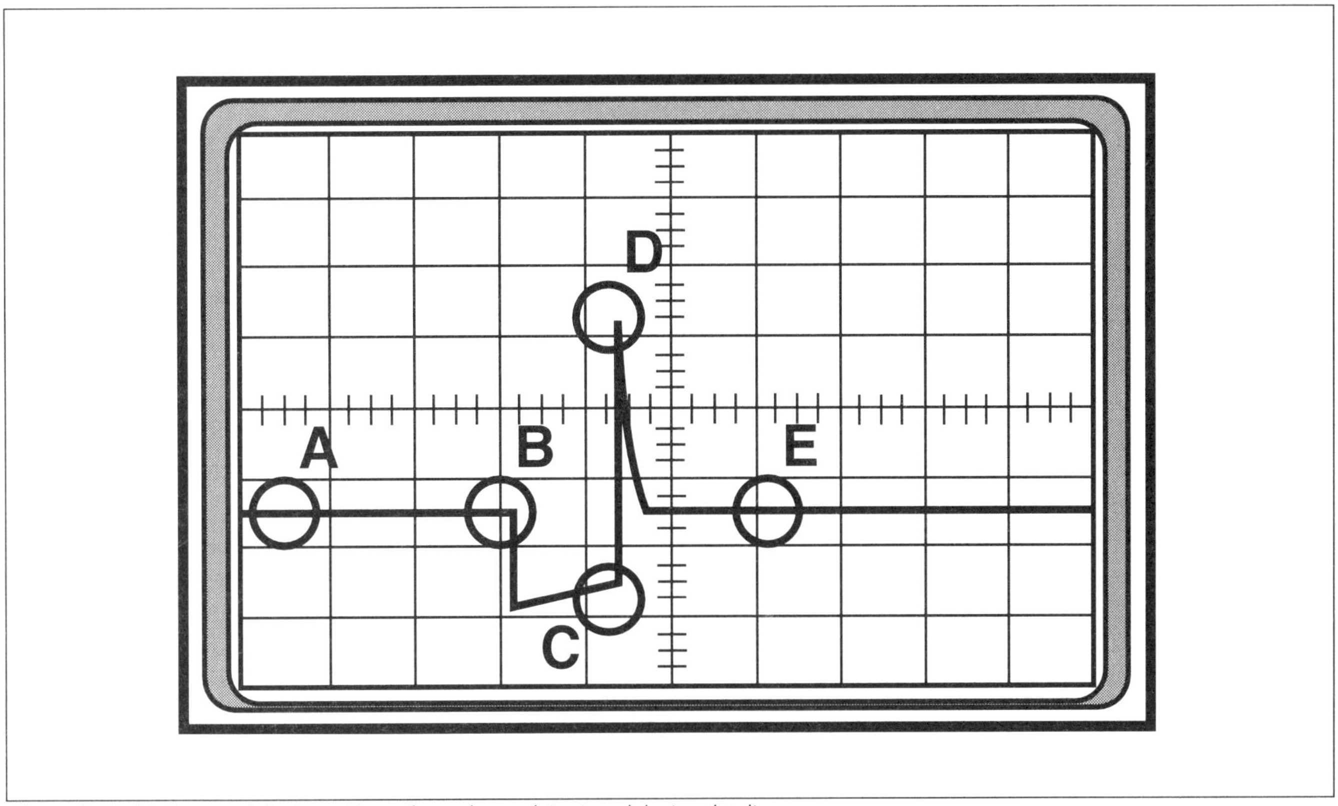

Figure 2-15. A lab scope injector pattern of a series resistor type injector circuit.

ic idle, engine stalling, and constant cooling fan operation.

- Intermittent IAT or ECT signals may cause constant hard starting, an erratic idle, engine stalling, surging, and engine stalling on deceleration.
- Coked throttle bodies and IAC conical valve may cause rough idle, hard starting, and engine stalling on deceleration.

Torque Converter Clutch (TCC) Solenoid Valve

The TCC solenoid valve is a PCM controlled valve inside the transmission, figure 2-17. When coolant temperature, vehicle speed, throttle position, engine load, and gear range are correct, the PCM energizes the TCC driver circuit. The TCC solenoid then engages the torque converter clutch by redirecting hydraulic pressure in the transmission or transaxle.

For the torque clutch to operate properly, the engine coolant temperature must reach about 150°F (66°C) and the vehicle must be at cruise (steady state throttle) above 40 mph. The Torque Converter Clutch (TCC) may show up in a data stream as: TCC on/off

TCC solenoid valve driveability problems

The following are examples of some of the driveability problems that can occur with torque converter clutch problems.

- False vehicle speed indication can cause premature or delayed locking of the torque converter clutch. Both problems can cause a loss of power and poor fuel economy.
- A shorted torque converter clutch circuit can cause a vehicle to stall when the car is started and then dropped into gear. This is caused by the TCC clutch being applied.

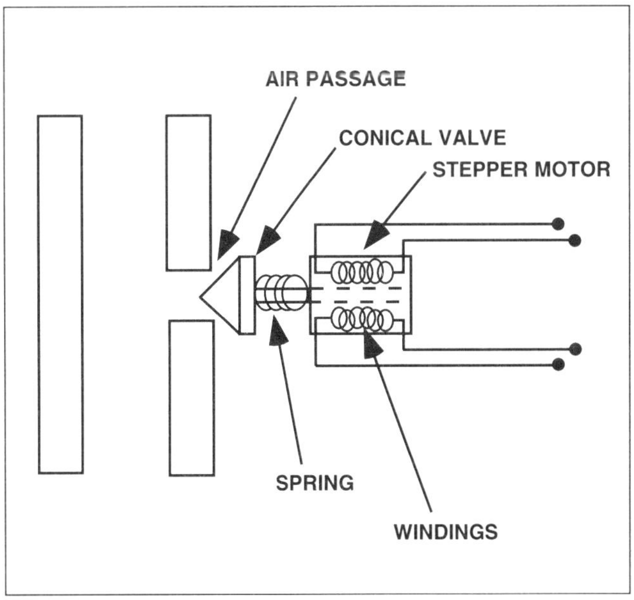

Figure 2-16. A stepper motor can make incremental adjustments in the idle-air passage to regulate idle speed.

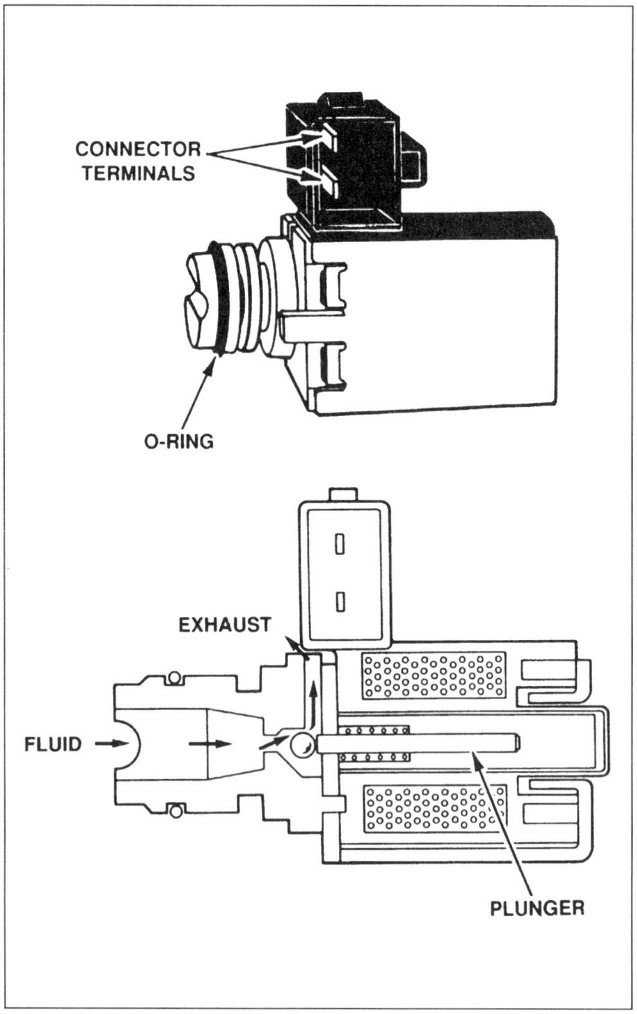

Figure 2-17. The Torque Converter Clutch solenoid applies the converter clutch by redirecting hydraulic pressure.

Ignition Control Module (ICM)

The ignition control module (ICM) processes a signal generated by the PCM to control electronic spark timing, figure 2-18. In computer timing mode, the PCM regulates the opening and closing of the primary circuit to control dwell and ignition timing advance. Note that portions of the ICM can act as an input device to the PCM by conditioning a crankshaft signal (CKP) before transmitting it to the PCM.

ICM ignition control information usually is not displayed as a data stream parameter. However, ignition problems, such as erratic ignition signal will set a DTC on some systems. The best way to look at this signal is with a engine analyzer or a lab scope.

ICM driveability problems and emission failures

The following are examples of some of the driveability problems and emission failures that can occur with ICM circuit problems.

- A faulty ICM can cause no-start, engine cut-out, surging hesitation, and excessive hydrocarbon emissions.
- An ICM that receives a faulty or inaccurate signal from the PCM can cause under advanced ignition timing, resulting in low power and poor fuel economy.
- An ICM that receives a faulty or inaccurate signal from the PCM can over advance the ignition timing, causing engine detonation, as well as high hydrocarbon and high NO_X exhaust emissions.

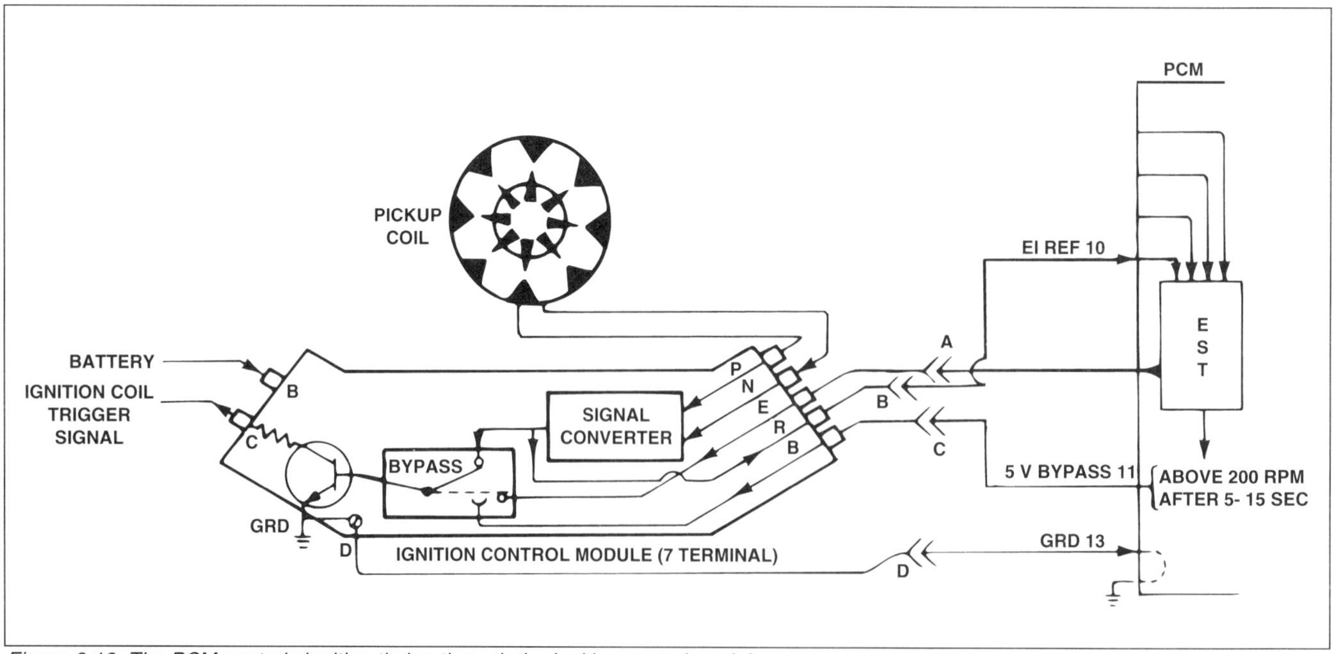

Figure 2-18. The PCM controls ignition timing through the ignition control module.

CIRCUIT TESTING USING SERIAL DATA

Using serial data to test PCM circuits can be of great value during driveability diagnosis, however, there are some items to remember. The data that is being read on the scan tool could actually be a default value that the PCM substitutes to compensate for possible circuit failures. Also, serial data transmitted by the PCM to the scan tool is an interpretation of what the PCM *thinks* it is seeing. The true readings may be different. You can confirm actual signal values by testing the circuit live with a DVOM, breakout box, or lab scope, depending on what you need to test. False data stream values may be caused by an internal PCM fault or a PCM ground circuit problem. The following are examples of using serial data to test and diagnosis driveability and intermittent problems:

- Thermistors: disconnect or short across thermistor circuit to check the maximum range of the sensor. For example, disconnect the ECT to create an open circuit. Temperature reading should drop to -40°F (-40°C). Install a jumper wire across the connector to create a short circuit. Temperature should go to a maximum reading, about 266°F (130°C).
- Create the opposite circuit problem to see if a DTC sets. For example, a P0122 code in memory tells you a TP sensor circuit voltage went low, indicating a short. To create an open circuit, disconnect the TP sensor and see if the PCM sets a P0123 (circuit high). If it does, then the circuit and PCM are operational and the problem is probably in the sensor.
- Intermittent problem testing: Wiggle, tap, heat up, or cool down a component or circuit to see if the serial data for that circuit changes.
- Testing the effect one circuit on another by manipulating the input signal. Manipulate the signal by disconnecting circuits or substituting values. Here are some examples:
- IAT, ECT, TP sensor, MAP, MAF, and HO2S signals' effect on injector pulse width.
- ECT, ACT, TP sensor signals' effect on Idle speed control.
- IAT, ECT, TP, MAP, and MAF signals' effect on ignition timing control.
- ECT, TP sensor, and EVP signals' effect on EGR control
- ECT and TP sensor signals' effect on canister purge
- VSS, TP sensor, ECT, and MAP signals' effect on torquo convert clutch operation.

Circuit Testing Operations

While scan tools are an important part of any diagnosis, once you locate a problem you must use either a DVOM or lab scope to accurately check a circuit. The following section covers circuit testing procedures and guidelines for using the proper test equipment.

Voltage

When using a DVOM to check voltage in and out of sensors, always check the voltage using the signal ground return at the sensor, rather than using an engine or battery ground, figure 2-19. Sensors are grounded directly through the PCM, rather than being connected directly to a chassis ground. This way sensors avoid noise interference. Sensors need a "clean" ground for reliable operation.

An open signal ground return will cause the PCM to see a high voltage on the sensor signal line. An example would be a TP sensor that always sends a wide open throttle (high voltage) signal to the PCM.

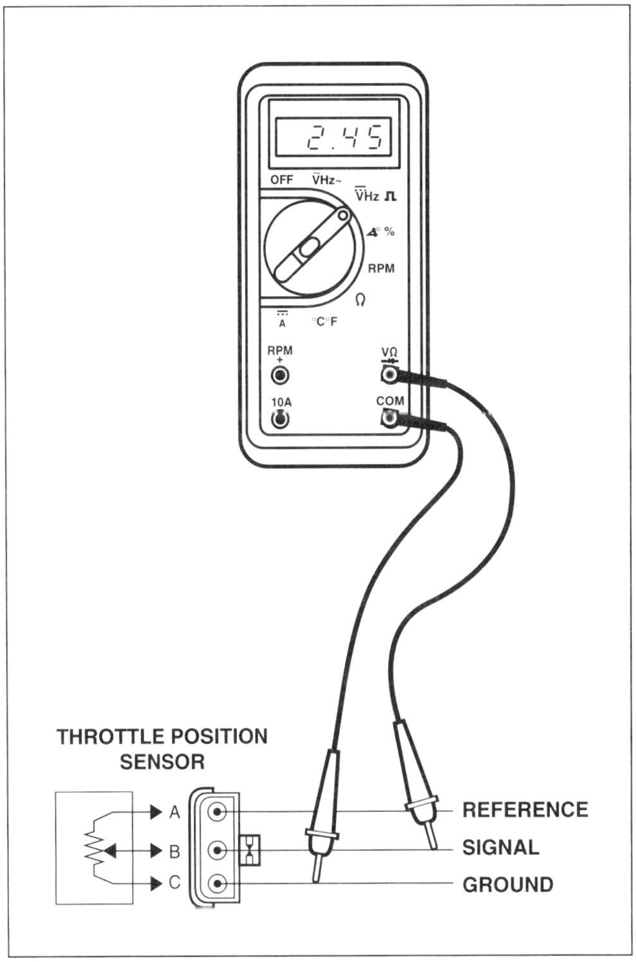

Figure 2-19. Checking voltage to a MAP sensor.

Resistance

Ohms law says that even very low resistance in an automotive computer circuit will cause sensors and actuators to work improperly because of low voltage. For example, an on-board PCM ignition feed circuit drawing 365 "milli" amps with a resistance in the ignition feed wire of 2.5 ohms, will cause a voltage supply drop of 1.5 volts. This voltage drop will cause severe driveability problems. One example of this may be a car that idles too high because the PCM monitors the battery voltage. If the supply voltage is low the PCM may raise the idle speed so the charging system could charge what the PCM thinks is a low battery.

To check resistance, make sure that the circuit to be tested is not under power. Place the leads across the circuit or component to be tested, figure 2-20. To read ohms, place the meter on auto-ranging or start at the higher scales and work down.

Voltage Drop

Checking voltage drop is one of the most important tests that a technician can perform on a circuit. Voltage drops can cause major driveability symptoms in on-board computer systems. A voltage drop on a PCM power ground can cause sensor voltage references to be higher than normal, throwing off the overall sensor calibration of the entire engine control system, figure 2-21.

Another example of a driveability symptom might be a car with an idle speed that continuously hunts. To start diagnosis, you connect the scan tool to check trouble codes and the idle smooths out. This is usually caused by a poor ground.

To check voltage drop, the circuit must be powered up and have current flowing. The circuit also must have the maximum amount of current flowing under normal conditions by which the circuit was designed.

Although there is no exact amount voltage drop that is considered acceptable, you should remember that low current circuits that draw milli amps will be affected by very small voltage drops. A good rule of thumb would be a drop of 0.2 volt or less. However, even this is too much for some circuits. A power ground circuit should have a voltage drop of no more than 0.1 volt.

Amperage

Too much amperage flow through a PCM actuator driver circuit can partially damage that circuit and cause severe driveability problems. Most PCM actuator components carry "milli" amps through their circuits. Using a ohmmeter and calculating amperage draw from resistance and voltage readings is not always accurate because the device under test does not carry the actual load it was designed to carry. Most actuator devices carry about 180 ma (12.6 volts at 70 ohms) to 500 ma (12.6 volts at 25 ohms), but there are always

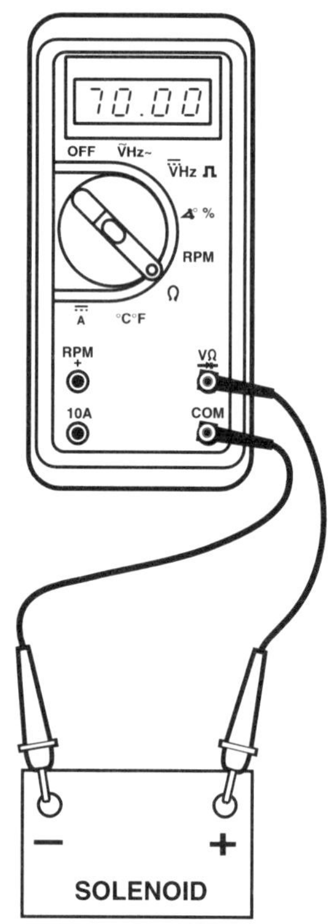

Figure 2-20. When checking resistance, the part must not be under power or you will probably destroy your meter.

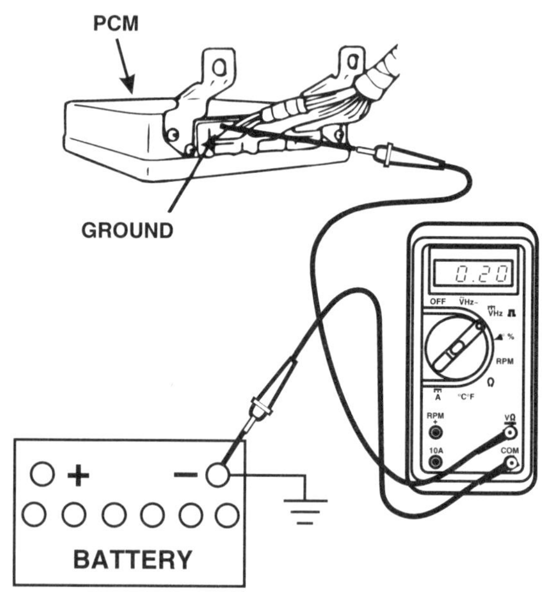

Figure 2-21. Checking voltage drop at the PCM ground connection.

exceptions to the rule. Fuel injectors may carry much more amperage through their circuit (as much as 8 amps depending on the type of injector).

To check amperage draw, the circuit must be powered up and have current flowing. Set your meter for amperage draw and connect it in series between the solenoid negative terminal and ground, or the actual driver circuit if you can energize it, figure 2-22. Start by checking amps first, then move down to "milli" amp scale. Leave the circuit energized for 1 to 2 minutes to check draw.

Remember, this test is for solenoids such as Canister Purge, EGR, Air Management, only. *Do not check fuel injectors in this manner.* Holding an injector on for any length of time destroys it.

AC Ripple

On-board automotive computers do not like to see AC ripples pass through the internal components. This effect can cause logic problems as well as many other types of driveability problems. For example, a bad alternator with a dropped diode can severely affect an automotive computer system.

To check for AC ripple voltage, switch your DVOM to AC and connect the black lead to a good ground and the red lead to the "BAT", or power, terminal on the back of the alternator (not the battery), figure 2-23. A good alternator should measure less than 0.5 volts AC with the engine running and the head lights on. A higher reading indicates damaged alternator diodes.

Frequency

Frequency is the number of times a signal repeats itself in one second. Frequency is measured in HERTZ. A signal that repeats itself 10 times a second, is operating at a frequency of 10 hertz.

Many automotive computer systems read the frequency of a signal instead of the voltage. Ford MAP sensors and AC Delco Mass Airflow Sensors are examples of sensors that produce this type of signal.

For example, a Ford EEC-IV MAP sensor has a 5 volt reference voltage applied to it. At a duty cycle of 50 percent (half of the time on and half of the time off), the DVOM will average the reading so you would see 2.5 volts. However, the number of times the signal switch-

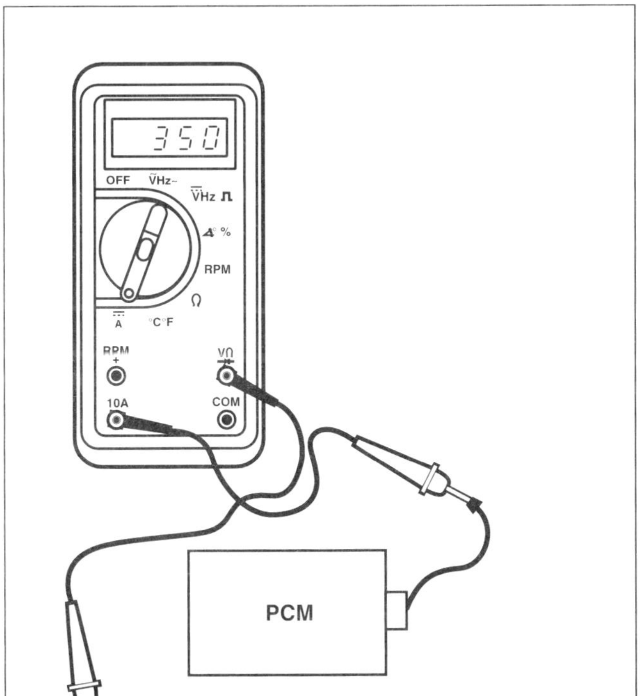

Figure 2-22. Checking amperage draw through a solenoid driver circuit.

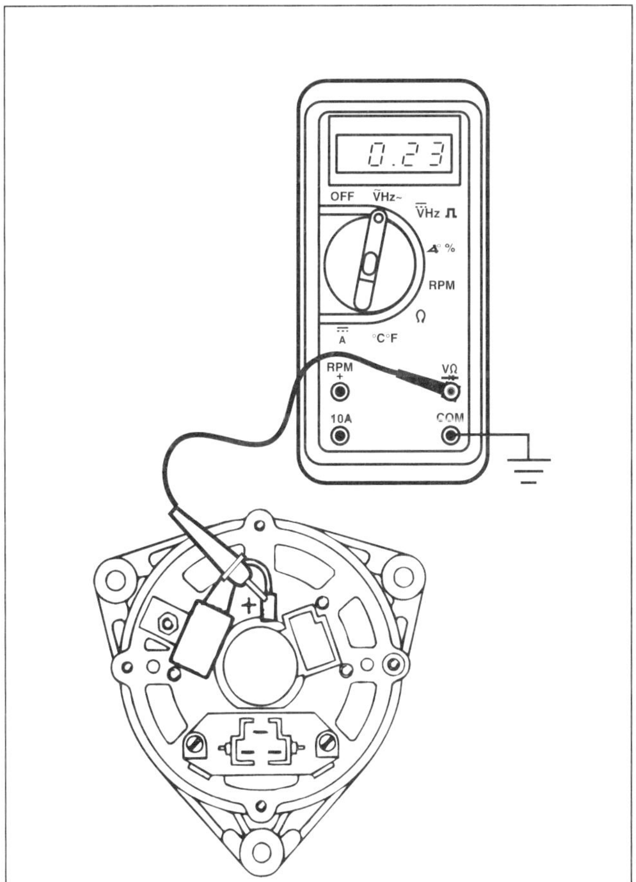

Figure 2-23. Checking for voltage ripple from an AC generator.

es on and off in one second will change depending on manifold vacuum. To accurately diagnose these signals, you must have a meter that can read frequency, figure 2-24.

Duty Cycle

Duty cycle is the percentage of time a digital signal is high verses low. When measuring duty cycle, one complete cycle is considered 100 percent. For a 5 volt signal at a 50 percent duty cycle, the voltage would read 2.5 volts.

For automotive applications, when dealing with digital waves, and especially with PCM outputs, we are concerned with the amount of time the signal is low, rather then high. This is because the low time is when the driving transistor is on, completing the circuit to ground.

You can measure duty cycle with a DVOM that has a duty cycle setting. Attach the red lead to the signal wire and the black lead to a good engine ground, figure 2-25.

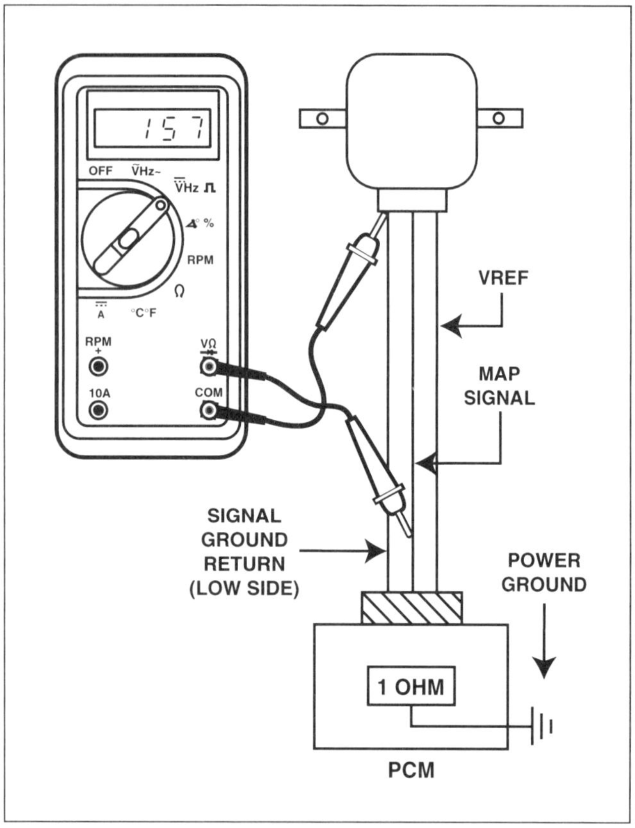

Figure 2-24. Checking the frequency of a MAP sensor.

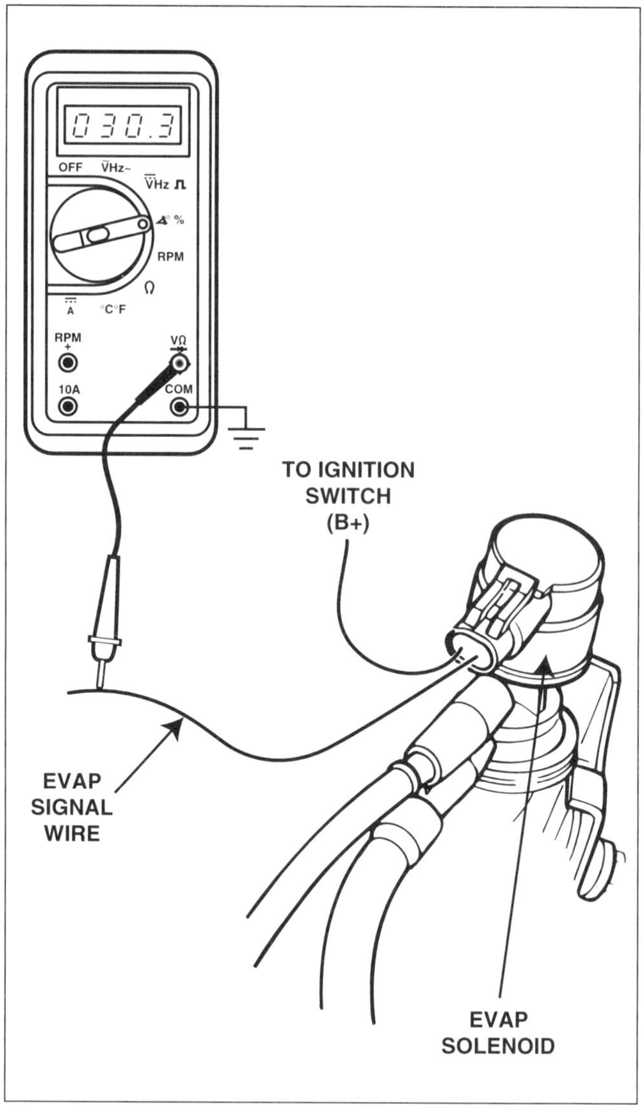

Figure 2-25. Checking the duty cycle of a canister purge solenoid.

1. A late-model engine computer typically *cannot* recognize that a:

   a. Signal is not being furnished

   b. Signal is improbable

   c. Signal is out of limits for too long

   d. Device is worn or in need of replacement

   Explanation: Answer d would be the only true answer before OBDII. However, now all cars meeting OBDII standards MUST be able to determine when a sensor or component is more than 40 percent worn or deteriorated.

2. Technician A says an analog sensor can develop dropout at some point of its range due to wear or age. This can cause an intermittent driveability problem.

   Technician B says if you operate a sensor through its full range, check the signal using a DVOM. First check at the sensor, then at the PCM harness connector.

   Who is right?

   a. A only

   b. B only

   c. Both A and B

   d. Neither A nor B

   Explanation: Signal dropout at one point, usually where most activity takes place, in a sensor's operating range is a common failure that can cause a driveability problem. Using a DVOM to check a sensor while you operate it through its entire range will pinpoint a dropout area. Checking at the PCM harness connector will eliminate any problems in the circuit. Therefore, both technicians are correct.

3. Technician A says that a faulty oxygen sensor on the composite vehicle may set a code P0172 (rich exhaust), but you should also check for problems in the manifold absolute pressure sensor, throttle position sensor, and other related circuits.

   Technician B says that a code P0172 (rich exhaust) may be caused by the system trying to compensate for a mechanical problem, such as leaking fuel injectors or a defective fuel pressure regulator.

   Who is right?

   a. A only

   b. B only

   c. Both A and B

   d. Neither A nor B

   Explanation: Code P0172 (rich exhaust) is a multiple circuit symptom code. Problems in other circuits, such as MAP or TP, that alter operating conditions can cause the code to set. A code P0172 can also result from a mechanical problem that allows an excessive amount of fuel to enter the engine, or misfire that prevents complete combustion. Therefore, both technicians are correct.

4. If a car with driveability problems runs better in the backup mode than it does normally, the problem probably lies in:

   a. The MAP sensor

   b. The PCM

   c. The catalytic converter

   d. The PROM

   Explanation: In the backup mode, the PCM operates according to a programmed set of instructions for fuel metering. If the engine runs better in the backup mode, a problem in the air intake or fuel metering signals is indicated. The MAP sensor is linked to both intake air and fuel metering, therefore, a is the correct answer.

5. Technician A says driveability symptom testing is relatively easy because problems set clear and recognizable trouble codes.

   Technician B says driveability symptom testing is generally more difficult because problems often set no, or misleading, trouble codes.

   Who is right?

   a. A only

   b. B only

   c. Both A and B

   d. Neither A nor B

   Explanation: This is a basic principle of troubleshooting control systems. Driveability complaints are typical of *any* vehicle. They are *not* just control system problems, and often set no trouble codes, so Technician B is correct.

6. Technician A says a rich exhaust trouble code is almost always an indication of a failing O2 sensor.

   Technician B says a rich exhaust trouble code is almost always an indication of a fuel-mixture imbalance.

   Who is right?

   a. A only

   b. B only

   c. Both A and B

   d. Neither A nor B

   Explanation: Although a rich exhaust code *can be* the result of a defective O2 sensor, it is *generally* an indication of a fuel-mixture imbalance rather than a hard sensor fault. O2S codes may be set by any condition that changes the air-fuel mixture, or causes the system to think the air-fuel mixture is out of range. Technician B is correct.

7. Technician A says a control system in open-loop will ignore exhaust oxygen sensor feedback signals.

   Technician B says a control system in closed-loop will ignore exhaust oxygen sensor feedback signals.

   Who is right?

   a. A only
   b. B only
   c. Both A and B
   d. Neither A nor B

   Explanation: This is a basic aspect of closed-loop operation. A system in closed-loop *uses* O2 sensor feedback signals to gauge the effects of its fuel metering and maintain the stoichiometric air-fuel ratio as closely as possible. By checking the effects of its last control actions, the system can correct itself, as necessary, and maintain very precise control. Technician A is correct.

8. Technician A says the resistance of an NTC-type thermistor decreases as temperature increases.

   Technician B says the resistance of a PTC-type thermistor decreases as temperature increases.

   Who is right?

   a. A only
   b. B only
   c. Both A and B
   d. Neither A nor B

   Explanation: The resistance of an NTC-type thermistor *decreases* as temperature increases, so Technician A is correct. The resistance of a PTC-type thermistor *increases* as temperature increases.

9. A typical piezoelectric knock sensor (KS) (signal generator) converts vibrations caused by engine knock into:

   a. Changes in the volummetric efficiency.
   b. Voltage signals at varying oscillations.
   c. A change in resistance.
   d. A change in capacitance.

   Explanation: Most piezoelectric knock sensors (KS) are tuned to respond to specific frequencies of the vibrations produced by engine knock. These vibrations are converted into voltage oscillations which are sent to the PCM or knock module, so b is correct. See the illustration of a Knock sensor (KS) taken with an engine under load.

10. The TPS signal below was captured during steady acceleration. All of the following is true about this signal **EXCEPT:**

   a. This TPS signal could cause the computer to go into clear flood mode.
   b. This TPS sensor signal could cause a hesitation and stumble on acceleration.
   c. This signal could cause erratic shifting on electronically controlled transmissions.
   d. This is a typical fault on a TPS sensor.

   Explanation: It is important to remember the word EXCEPT. Answer a is the only condition which would not occur. For the PCM to go into a clear flood mode, the TPS sensor voltage must exceed 80 percent of full throttle or 3.7 volts with the RPM's below 400.

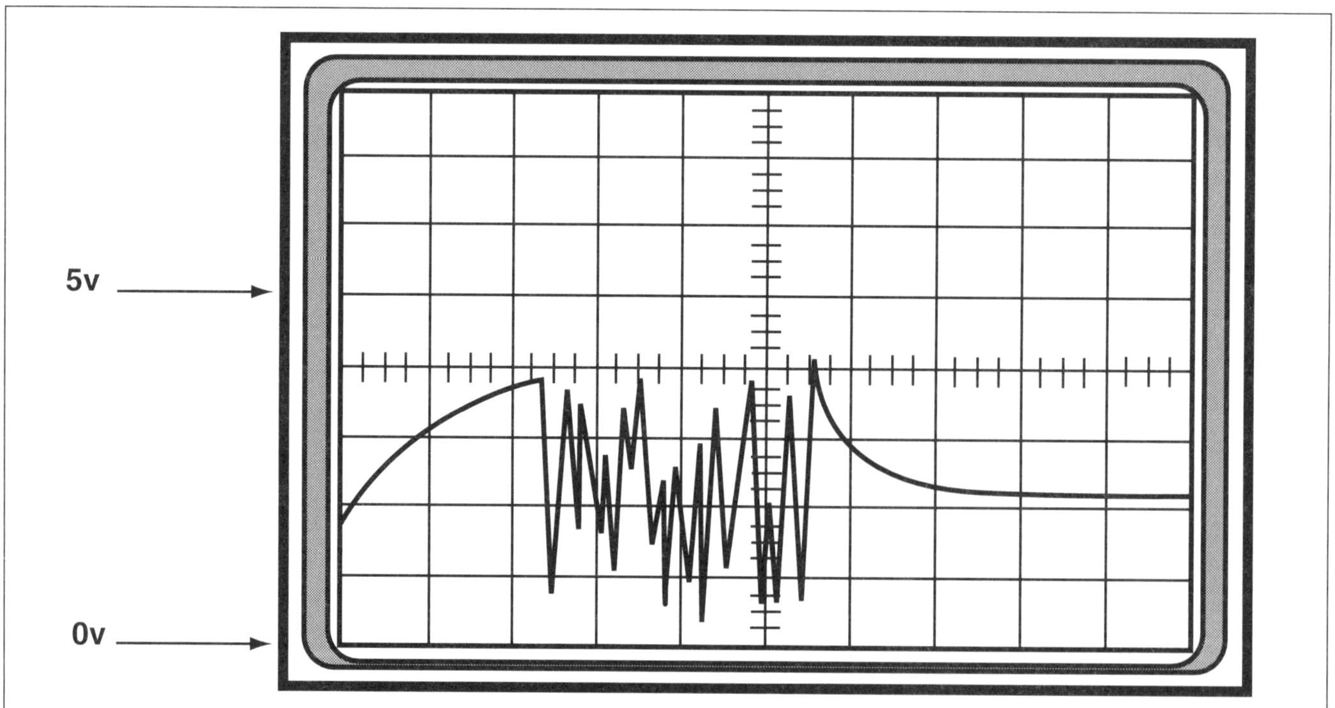

Illustration for question 10.

Chapter Three

IGNITION SYSTEMS DIAGNOSIS

In this chapter we discuss basic electrical schematic diagrams and the flow of power through the control systems circuits. We also examine oscilloscope patterns, and how they are a useful diagnostic aid for locating problems in the primary and secondary ignition systems. The chapter concludes with an explanation of radio frequency interference (RFI) and electro-magnetic interference (EMI), and how these noise transmissions can effect the integrity of the electronic control system.

GENERAL ELECTRICAL DIAGNOSIS

An electrical schematic, or wiring diagram, is an indispensable tool for troubleshooting vehicle electronics. A wiring diagram tells you how power flow is directed through the system, the functional nature of individual components, and how various components are linked through the circuitry. Without this knowledge, you are simply guessing at how the system functions. Vehicle manufacturers provide wiring diagrams in their service manuals, and they are also available from a variety of aftermarket sources. Locating an accurate wiring diagram should always be the first step in electrical diagnosis.

Electrical Diagrams

For the L1 test, ASE designed a composite vehicle with a unique engine control system. Compared to real world applications, the composite vehicle system is simplified and straight forward. All pertinent information, including a wiring diagram, is provided in the Composite Vehicle Preparation/Reference Booklet.

Not all of the L1 test questions pertain to the composite vehicle. However, if you have a good understanding of composite vehicle functions, you should be able to correctly answer all test questions based on logic and deductive reasoning. For your convenience, a reproduction of the composite vehicle wiring diagram is included as figure 3-1 of this book.

In addition to providing a layout of low-voltage circuits, the composite vehicle diagram also serves as a pin-out chart by numbering and identifying pin connections at the PCM. Noticeably absent from the schematic are the charging system, ignition switch, and secondary ignition circuits. Assume charging and ignition switch systems are in good condition if battery voltage is available at pins 11 and 12 of the PCM. Use the standard diagnostic procedures discussed in this chapter for isolating problems in the secondary ignition circuits.

Standard industry symbols identify the electrical and electronic components used on the system. These symbols are internationally recognized, but may vary slightly between different manufactures and by their country of origin. Since they are universally recognized, symbols provide a quick reference to the design of individual components. Figure 3-2 provides descriptions of the symbols used on the composite vehicle diagram.

Control System Power Flow

In discussing power flow through the electronic control system we will use the composite vehicle as an example. Although this is a basic and simplified system, the principles that apply here also apply to most automotive control systems. The only major differences between most systems are the number of controlled circuits, the design characteristics of individual components, and the arrangement of PCM terminal connections, or pins.

Internal PCM functions

Battery, or system, voltage is supplied to the PCM at two points, pins 11 and 12. To understand the need for two power sources, you must be familiar with the basic functions of a computer system, figure 3-3. All automotive computers perform four functions:

1. **Input** — Information on engine operating conditions provided to the computer, or PCM, as voltage signals from system sensors.
2. **Processing** — The internal function that converts input voltage signals into combinations of binary numbers (1's and 0's). The PCM then compares the binary information to instructions and other

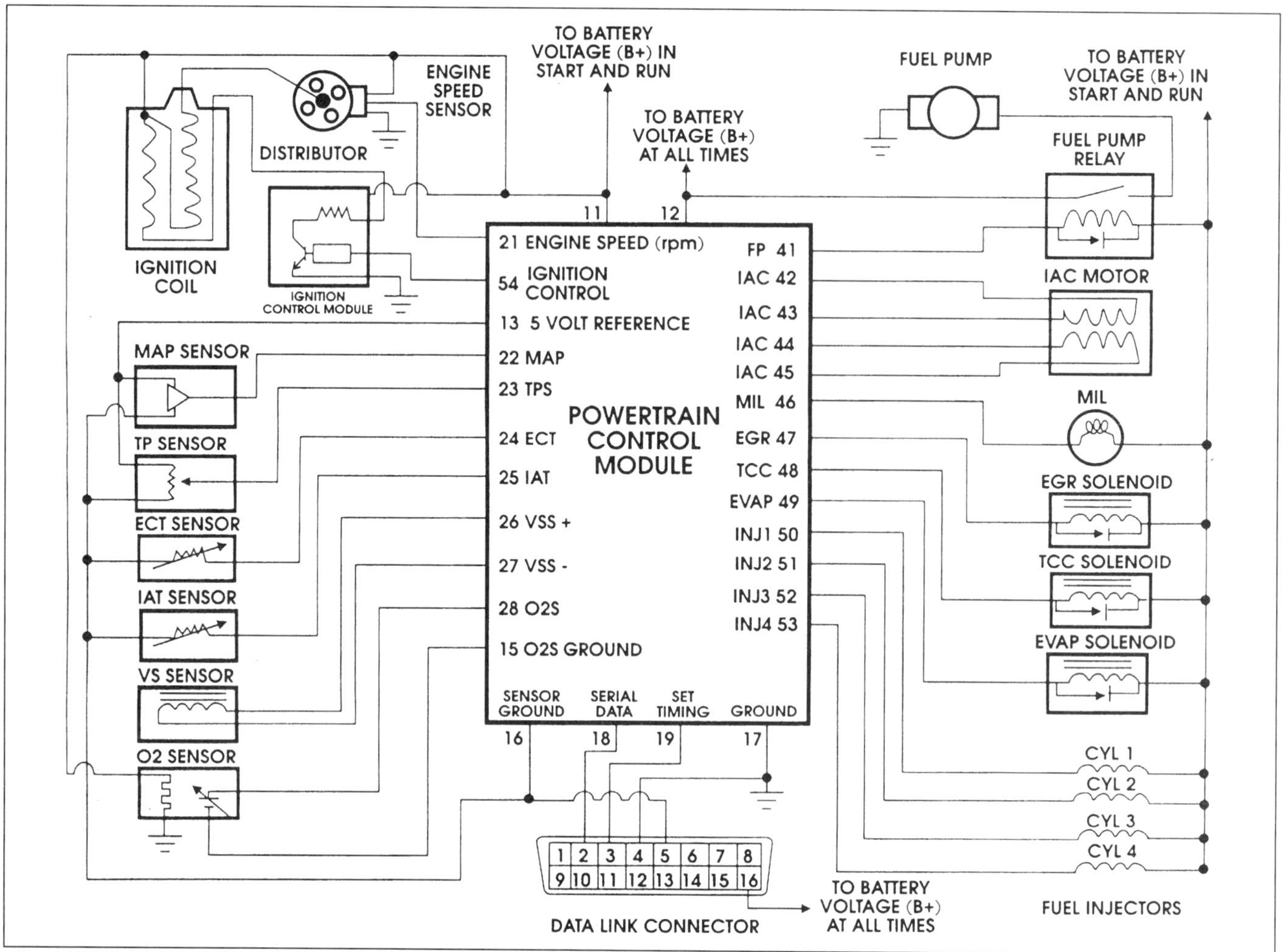

Figure 3-1. Composite vehicle electrical schematic.

programmed data to make logical decisions and send output commands to control other devices in the system.

3. **Storage** — All computers have some kind of information storage capacity. This electronic memory stores the computer operating instructions, or program. In addition, some computer systems also store certain input signals for later reference. These computers "remember" previous operating conditions and adapt output commands to the "learned" characteristics of vehicle operation. This feature, known as adaptive strategy, allows the system to compensate for wear, climatic conditions, driving style, and other variables.
4. **Output** — After receiving an input signal, processing it, and comparing it to information stored in memory, the computer transmits a command to output devices that control engine operation. Output devices, such as solenoids and relays, are actuators that convert electrical signals to mechanical motion to obtain a predetermined response and regulate engine operation. The computer may also send output information to display devices such as electronic instrument panels.

On the composite vehicle, pin 12 provides the power supply for the PCM storage circuits. Internal memory circuits can be further broken down into two categories: volatile and non-volatile. Volatile memory is temporary and requires a constant power supply to retain stored information, information stored here is lost if the battery is disconnected. Information that changes during vehicle operation, such as adaptive strategies and diagnostic trouble codes, are stored in volatile memory. Information in the non-volatile memory is permanent and is not dependant on battery power. Basic operating program information and look-up tables, which are not needed unless the engine is running, are stored in the non-volatile memory or PROM.

Switching the ignition key to the start or run position directs an additional source of system voltage to pin 11 of the PCM. This provides the necessary power to operate the processing circuits. Power applied here is subsequently directed to the input sensors through a

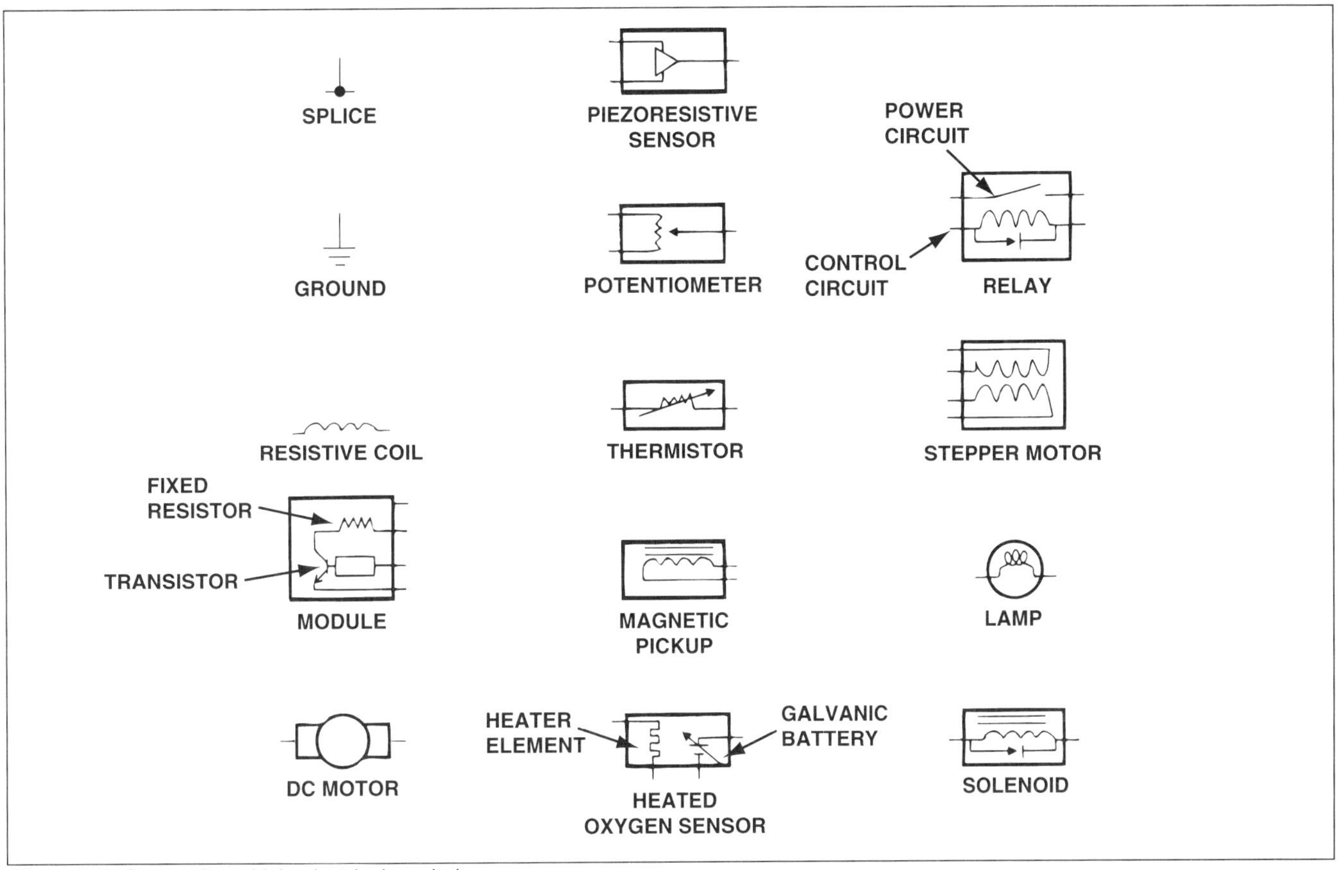

Figure 3-2. Composite vehicle electrical symbols.

voltage regulator as a low-voltage reference signal, and as an output voltage signal to the actuators. With the exception of ignition and fuel pump circuits, most input sensors operate on low voltages that seldom exceed 5 volts. However, the internal processor of the PCM requires system voltage. Most control systems will suspend operations and revert to a fail-safe, or limp-home, strategy should supply voltage fall below about 10 volts.

Input signal processing

To process information from an input sensor, the PCM applies a low-voltage reference signal, usually about 5 volts, to the sensor. Most two-wire sensors have the power source and a fixed resistor inside the PCM. The PCM reads the voltage after the fixed resistor. Current flows through the fixed resistor, and then through the variable resistor in the sensor, and back to ground. As the resistor in the sensor changes, the voltage in the PCM changes. As the sensor resistance decreases, the voltage drop decreases, and the read voltage decreases.

A three-wire sensor is usually a potentiometer. A 5-volt source is applied to one end of a resistor, and the other end is grounded. The voltage drop is proportional to the position of a sweep arm that sends the signal back to the PCM.

Because the PCM operates on low voltage signals, even small changes in resistance anywhere along a circuit can effect system performance. Many driveability complaints and control system malfunctions are a direct result of excessive resistance on the monitored circuits. Poor (loose, dirty, or corroded) connections and damaged (frayed, pinched, cracked, or broken) wiring and insulation can all be a source of high circuit resistance. As explained in Chapter 2, the most effec-

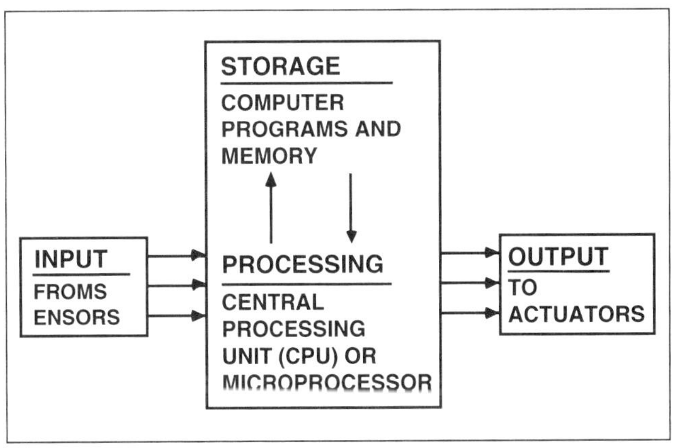

Figure 3-3. All computers perform 4 basic functions.

tive way of locating a source of high resistance on a computer-controlled circuit is voltage drop testing.

Output signal processing

The PCM switches output circuits open and closed to command system actuators to turn off and on. Power to most actuators is provided through the ignition switch circuit, and the PCM regulates voltage on the ground side. Actuators respond to PCM commands and, in turn, control other circuits or convert electrical energy into mechanical work. PCM circuits that command actuators are often referred to as driver circuits. Typical output devices include:

- Solenoids
- Relays
- Stepper motors.

Almost all actuators in engine control systems contain some kind of induction coil. See Chapter 2 for more detailed information on actuator design and function.

Every output device controlled by the PCM has a minimum resistance specification, and this is an important consideration when servicing the output side of the system. Internal resistance in an actuator limits the amount of current that is allowed to flow through the output control circuit. If an actuator is shorted, current in excess of the safe maximum limit can flow to the PCM and cause severe damage. Typically, a controlled output will allow no more than 0.75 ampere (750 milliampere) to flow through the circuit. Shorted actuators are a common cause of PCM failure. If not replaced, a defective actuator can quickly destroy a new PCM. Always check output circuits for shorts and low resistance when replacing a failed PCM.

IGNITION SYSTEM DIAGNOSIS

Although most ignition systems are incorporated into the engine management system, ignition failures generally do not set trouble codes and ignition malfunctions cannot be detected on the serial data stream. The steps taken to isolate the source of an ignition system failure vary by vehicle and symptom. The material presented here follows a general procedure for dealing with a no-start condition, and takes into consideration different component design possibilities.

Keep in mind, electronic ignition systems contain delicate, expensive components a voltage surge can destroy. The high-voltage surge that results from simply making or breaking a connection can cause damage. When working on the ignition system, take these precautions to avoid damage and injury:

1. Unless the procedure recommended by the manufacturer specifically states otherwise, switch the ignition off or disconnect the battery ground cable before separating any ignition system wiring connections.
2. Do not touch, or short to ground, any exposed wiring or connectors while the engine is cranking or running.
3. Do not short the primary circuit to ground without resistance or the high current will damage wiring and destroy electronic components.
5. Do not create a secondary voltage arc near the battery or fuel system components.
6. When testing the secondary voltage delivered by the coil, follow the exact procedure specified by the manufacturer.

Basic Spark Check

For a no-start condition, verify that secondary voltage is available from the coil by performing a spark test. On some systems, you can disconnect a spark plug cable from the plug and hold the disconnected end about 1/4 inch (6 mm) from a good ground. Then, watch for a spark while cranking the engine. A bright blue spark should be clearly visible. This method can create high open-circuit voltage and damage electronic components, and should only be used when recommended by the vehicle manufacturer. Use of a spark plug simulator or a special test light is not only safer, it is often the preferred method.

A spark plug simulator is basically a plug without a side electrode and a grounding clamp attached to it, figure 3-4. Connect the secondary coil lead, or plug cable, to the simulator and clamp it firmly on a good ground. Cranking the engine should produce an arc from the center electrode to the side of the simulator. Test lights designed to handle secondary voltage can be installed in-line on any of the high-tension cables.

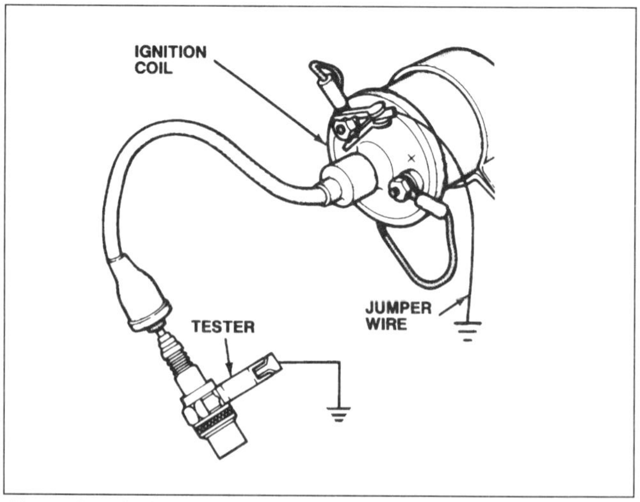

Figure 3-4. A plug simulator prevents high open-circuit voltage when performing a spark check.

Voltage passing through the tool will illuminate the lamp. Test lights can be used both while cranking, or running the engine.

On some Ford ignition systems, the magnetic pickup in the distributor is located directly under one of the high-tension terminals in the cap. Opening the circuit to this plug can allow the rotor to cross fire to the pick-up coil and damage the assembly. Do not remove the following cables when performing a spark test on a Ford product:

- Inline 4-cylinder engines — Nos. 1 and 3
- Inline 6-cylinder engines — Nos. 3 and 5
- V6 engines — Nos. 1 and 4
- V8 engines — Nos. 1 and 8.

A lack of secondary voltage can be caused by a defective coil, or a number of other problems in the primary circuit.

Ignition Coil Testing

A series of tests can be performed on an ignition coil using a voltmeter, ohmmeter, and ammeter. Most vehicle manufacturers provide specifications for coil testing, have these on hand before you begin. The coil must be at normal operating temperature for accurate test results. Begin by making these preliminary checks:

- Be sure all connections are clean and tight
- Look for a cracked or burned coil tower
- Look for a cracked or damaged coil housing.

Winding resistance tests

The first test checks the resistance of the coil windings using an ohmmeter. Disconnect the wires from the primary coil terminals, and the high tension lead from the coil tower. Set the ohmmeter to the low scale, and test as follows:

1. Connect one ohmmeter lead to the coil positive, or battery primary terminal.
2. Connect the other ohmmeter lead to the coil negative, or distributor primary terminal, figure 3-5.
3. Compare the ohmmeter reading to specifications. Typically, primary winding resistance will be in the 0.50 to 2.00 ohm range.

To test secondary winding resistance, set the ohmmeter to the highest scale and proceed as follows:

1. Connect one ohmmeter lead to the coil tower secondary terminal.
2. Connect the other ohmmeter lead to one of the primary terminals, figure 3-6.
3. Compare the ohmmeter readings to specifications. It is important to look up the correct specifications

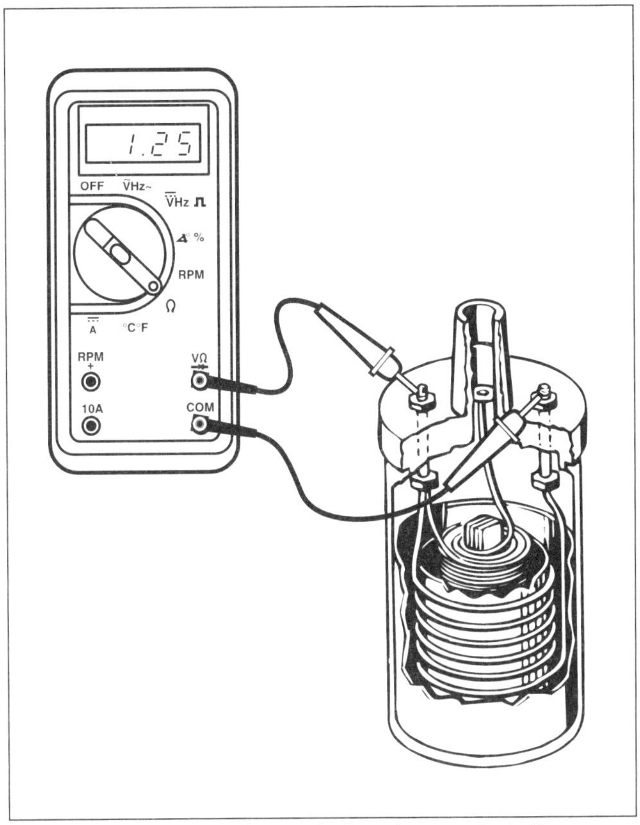

Figure 3-5. Checking resistance in the primary coil winding.

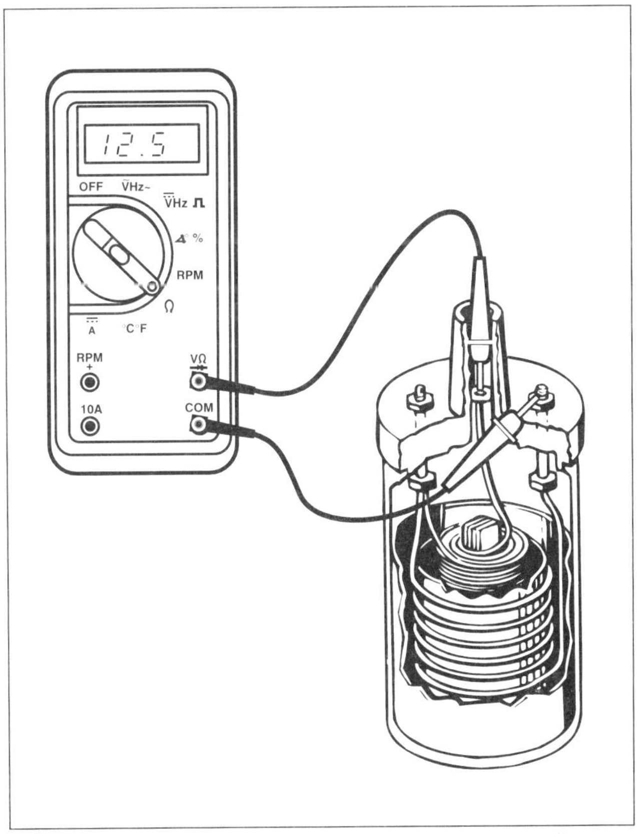

Figure 3-6. Checking resistance in the secondary coil winding.

for each coil. Some coils connect the secondary ground to the primary circuit, but others have a separate ground for the secondary circuit. If your are not using the correct specifications, you could have the wrong test results for secondary winding resistance, or you could short to ground through the case.

Next, test the coil windings for shorts. Set the ohmmeter on the lowest scale, and proceed as follows:

1. Connect one ohmmeter lead to either primary terminal.
2. Touch the other ohmmeter lead to the metal case of the coil. The meter should indicate infinite resistance. Any other reading may indicate the coil windings are shorted to the case and the coil must be replaced. In DIS systems, shorted primary coil windings can also damage the ignition module driver for that coil.

Current draw test

You can check the current draw of a coil on a running engine using an ammeter. Switch the ignition off when connecting the meter. If the ammeter has an inductive pickup, clamp the pickup into the positive primary coil lead. Use the following procedure for an ammeter without an inductive pickup:

1. Disconnect the positive primary wire from the coil.
2. Connect the ammeter positive lead to the positive primary wire, and connect the ammeter negative lead to the positive primary terminal at the coil, figure 3-7.
3. Check the specific instructions for the vehicle being tested.
4. With the engine running, the ammeter displays current draw.

No current draw indicates an open circuit. Look for broken or disconnected primary wiring.

Low coil current draw indicates high resistance. This can be caused by a discharged battery, damaged primary wiring, a defective coil winding, and loose or corroded connections.

High current draw indicates low resistance that often results from a shorted coil winding. A replacement coil that does not meet original equipment specifications can also draw too much current.

Polarity test

If the coil polarity is reversed, the engine will start and run but required voltage can increase by 20 to 40 percent. Under some operating conditions, this can cause misfiring that leads to an increase in exhaust emissions. If you cannot determine coil polarity from the

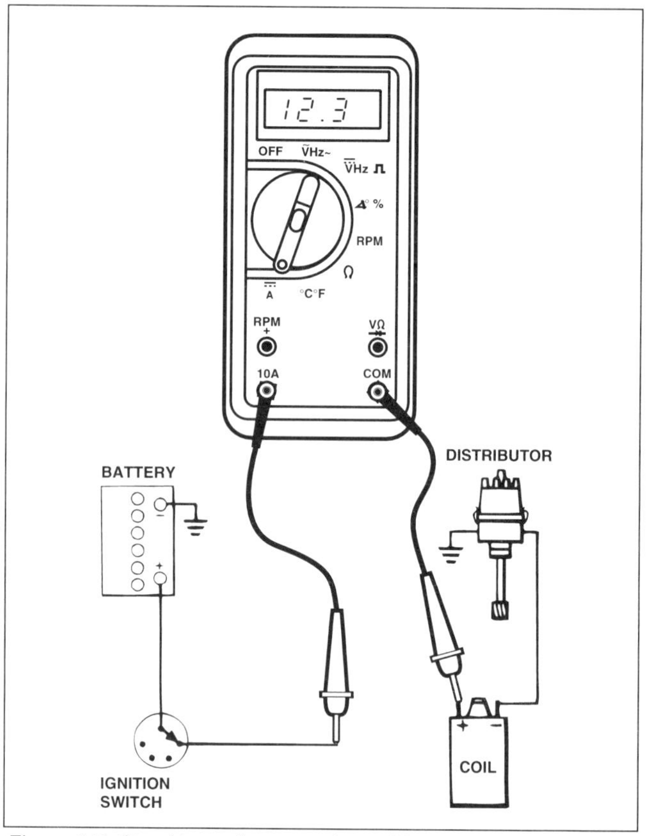

Figure 3-7. Checking coil current draw with an ammeter.

markings at the coil primary terminals, check polarity with an oscilloscope. Reversed polarity results in an inverted, or upside-down, scope pattern. Scope patterns will be discussed later in this chapter.

Primary Circuit Testing

Primary circuit voltage has a direct effect on secondary circuit voltage. The loss of a single volt in the primary circuit can reduce secondary circuit voltage by as much as 10 kilovolt (10,000 volts). Common causes of primary circuit voltage loss include: high circuit resistance, insufficient source voltage from the battery, and low charging system output.

Sources of high primary resistance include:

- Loose, corroded, or damaged wiring connections
- An incorrect or defective coil
- A poor ground at the ignition module.

Low source voltage can be caused by the following:

- Excessive starter motor current draw
- Low charging voltage
- A discharged battery.

Primary circuit voltmeter tests

First, make sure the battery has the correct performance rating and is fully charged. Then, disable the ignition system following the specific procedures provided by the manufacturer. Next, check for:

- Available voltage while cranking
- Voltage-drop across the battery ground.

To test voltage available from the battery while cranking, you can monitor the data stream with a scan tool, or:

1. Connect the voltmeter positive lead to the positive battery terminal, not to the cable connector. Connect the voltmeter negative lead to a good ground.
2. Switch the ignition key on and take a voltmeter reading while the engine is cranking.

A reading of ten volts or more indicates the battery is in good condition. Low readings can be caused by a voltage drop across the battery ground cable, excessive starter motor current draw, or incorrect charging system output.

To test voltage drop across the battery ground cable:

1. Connect the voltmeter positive lead to ground and the negative lead to the negative battery terminal, not to the cable connector, figure 3-8.
2. Crank the engine with the starter and take a voltmeter reading.

If the reading is 0.2 volt or less, the battery ground connection and cable are in good condition. When readings exceed 0.2 volt, clean and tighten the battery cable connections and terminals. Repeat the test; replace the ground cable if readings remain high.

Additional ignition test procedures, and the equipment used vary by system. Accurate specifications and instructions from the manufacturer are required. The following paragraphs discuss general procedures for testing a variety of electronic components.

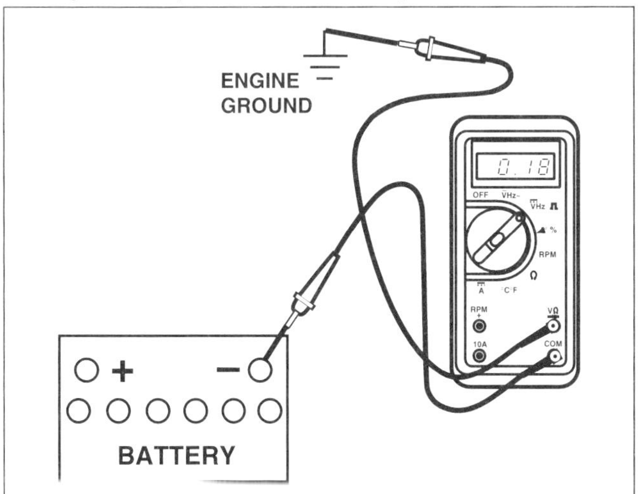

Figure 3-8. Voltage drop at the battery ground cable should be less than 0.2 volt.

Electronic Component Testing

Primary ignition failures often originate from problems in the:

- Primary wiring and connectors
- Distributor signal generator
- Ignition control module.

Primary wiring and connectors can be a source of high circuit resistance, as well as unwanted opens and grounds. Simple voltmeter and ohmmeter tests will generally isolate the source of the failure. The first step is to check the power supply at the PCM connector with a voltmeter. On the composite vehicle, battery voltage should be available at pin 12 at all times, and at pin 11 when the ignition switch is in the start or on position.

If correct voltage is present at the PCM, work your way down the circuit checking for dropped voltage. On the composite vehicle, voltage drop checks would be made in the following order:

- Ignition control module (ICM)
- Distributor (engine speed sensor)
- Ignition coil
- Oxygen sensor heater.

When voltage drop indicates high or low resistance, disconnect power and verify your findings with an ohmmeter. Most manufacturers provide resistance specifications for components and circuits. Many problems can easily be solved by cleaning connectors, related grounds, or repairing damaged wiring.

In general, an electronic ignition system functions as follows: The distributor signal generator sends a signal to the PCM. The PCM modifies the timing and sends the signal to the ignition control module to toggle the primary coil circuit and fire the spark plugs. Some signals are generated by a pickup coil or Hall-effect switch mounted in the distributor housing. Direct, or distributorless ignition systems use a crankshaft position (CKP) sensor to generate an input signal to the ICM, figure 3-9. The ICM provides the output signal to toggle the primary circuit and initiate firing. The system may use an additional camshaft position (CMP) sensor to determine cylinder phasing and achieve more accurate control of plug firing and timing advance, figure 3-10.

Methods to determine whether an ICM is good or defective varies between systems. Procedures outlined by the vehicle manufacturer must be followed. In general, eliminate all other possibilities before condemning the module. Verify power is available to the module, and check for voltage drop across connections on the power and ground circuits. Since modules

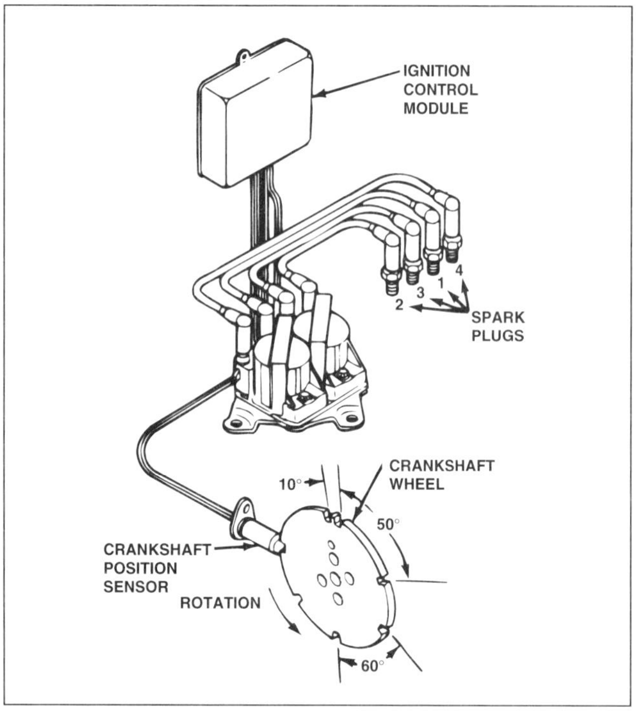

Figure 3-9. A crankshaft position sensor provides the input used to initiate plug firing on a direct ignition system.

provide the primary circuit ground connection, many problems can be solved by checking and cleaning the ground.

For any ignition system to operate properly, the following conditions must be met:

- Battery cranking voltage must be over 9 volts
- Primary wiring must be in good condition, and all connections need to be clean and tight
- Spark must be available at the spark plugs while the engine is cranking.

The air gap between the pickup coil and trigger wheel is adjustable on some systems. Gap has no effect on the dwell period; dwell is determined by the control module. The air gap must be set to a specific clearance when a new pickup unit is installed. During use, the air gap should not change. However, it should be

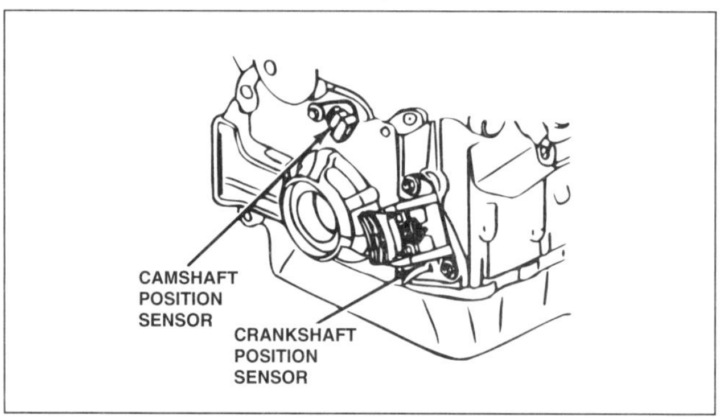

Figure 3-10. Some systems use an additional sensor on the camshaft to properly phase the firing sequence.

checked before performing troubleshooting tests.

Air gap check and adjustment

Distributor bushings can wear from simple rotation on a high-mileage distributor. Also, the breaker plate and any advance mechanism can wear and become loose. All of these conditions can change the air gap between the trigger wheel and the pickup coil.

Measure air gap using a nonmagnetic (brass) feeler gauge, figure 3-11. The pickup coil magnet will attract a steel gauge, causing an inaccurate adjustment. Air gap differs between manufacturers and models, so you must have the correct specifications. To check and adjust air gap:

1. Rotate the engine to align one trigger wheel tooth with the pickup coil.
2. Place a nonmagnetic feeler gauge of the specified thickness between the wheel tooth and the pickup coil. The feeler gauge should just barely contact on both sides.
3. If adjustment is necessary, loosen the lock screws and shift the pickup plate. Tighten the lock screws and remove the feeler gauge.
4. Verify the adjustment using a "no-go" feeler gauge.
5. If the distributor has a vacuum advance unit, apply vacuum to the diaphragm as you watch the breaker plate move through its full travel. Be sure the pickup coil does not hit the trigger wheel tooth. Release vacuum and recheck the air gap. If the pickup plate is loose, overhaul or replace the distributor.

Signal generator service

Distributor-type ignition systems use a solid-state switch to make and break the coil primary circuit. Several different switch designs are in use. Two common switching devices, the pulse generator and the metal

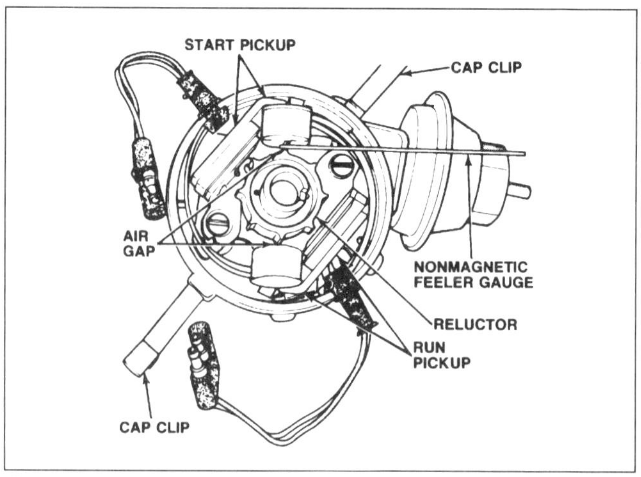

Figure 3-11. Checking air gap on a distributor with a permanent magnet pickup coil.

detector, consist of a stationary electromagnetic pickup coil and a rotating trigger wheel. Another type, the Hall-effect switch, uses a semiconductor and a permanent magnet to produce a voltage pulse and provide the primary trigger signal. A few systems use an optical sensor to cycle the circuit. On an optical sensor, a slotted shutter wheel passes between a light-emitting diode (LED) and a phototransistor to toggle the signal on and off. Refer to Chapter 2 for more detail on these systems.

Distributorless, or direct, ignition systems trigger firing based on input signals from crankshaft position and camshaft position sensors. Most direct ignition systems use permanent magnet-type sensors, however, some use Hall-effect switches. ASE does not specify which type of pickup is used on the composite vehicle.

Pick-up coil testing

When troubleshooting tests point to the pickup coil, check coil resistance with an ohmmeter. All pickup coils operate at a specific resistance. Specifications, test points, and procedures will vary, you must have accurate information from the manufacturer.

In general, measurements are taken by connecting the ohmmeter leads to opposite sides of the coil, figure 3-12. If the reading is not within specifications, replace the pickup coil. To check for a grounded coil, connect one ohmmeter lead to ground and touch the other ohmmeter lead alternately to each of the pickup coil connectors. The ohmmeter should show infinite resistance at all test points. If not, replace the pickup.

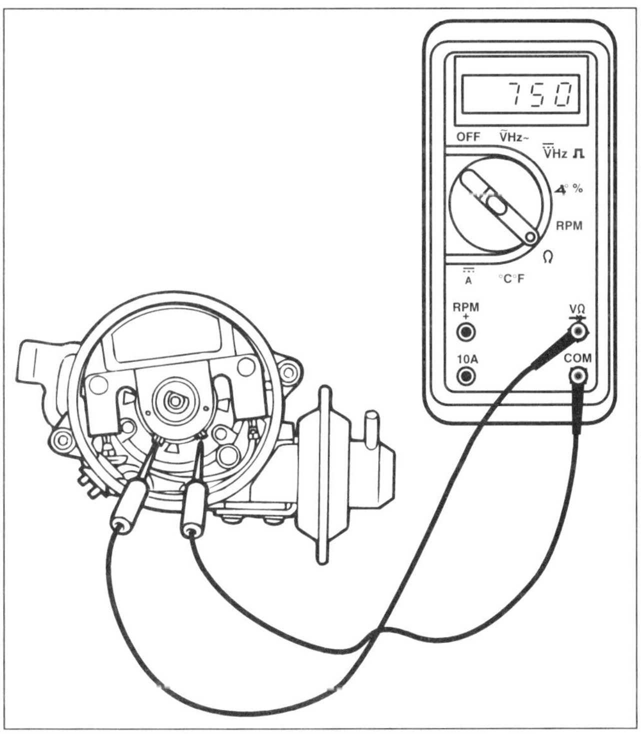

Figure 3-12. Using an ohmmeter to check pickup coil resistance.

Magnetic pickup sensor testing

You can check all permanent magnet position sensors for internal circuit problems by measuring resistance with a digital volt-ohmmeter (DVOM). You can monitor most sensor voltage signals with a DVOM, scan tool, or oscilloscope while the engine is running, figure 3-13.

Magnetic sensor output voltage is related to engine speed. Typically, it ranges between about 500 millivolts at cranking speed to approximately 100 volts at high rpm. When diagnosing an engine that cranks but will not run, the sensor output should exceed the minimum specified output while cranking. Minimum output will vary according to ambient temperature, oil viscosity, condition of the starting system, and the quality of the test equipment being used.

Since position sensors operate at low-voltage, they produce a voltage trace that can be monitored on a laboratory oscilloscope. Compare the actual scope trace to that of a known good pattern to determine sensor condition.

Pickup sensors are not adjustable. However, they must be properly aligned and installed with a precise gap between the interrupter blades and sensor to function properly. Misalignment, incorrect gap, damage to the interrupter rings, or accumulated debris can cause the engine to sag, stumble, hesitate, or diesel. Sensor failure can prevent the engine from starting, or may cause it do die immediately after startup. Engine misfire, stumble, or intermittent no-start conditions can result from sensor damage due to road debris or vibration.

Hall-Effect switch test

Hall-effect switches are used in conjunction with a magnetic pickup coil. The Hall-effect switch generates high signal voltage when the magnetic field is strong, and low signal voltage when the field is weak, figure 3-14. The signal will produce a digital square

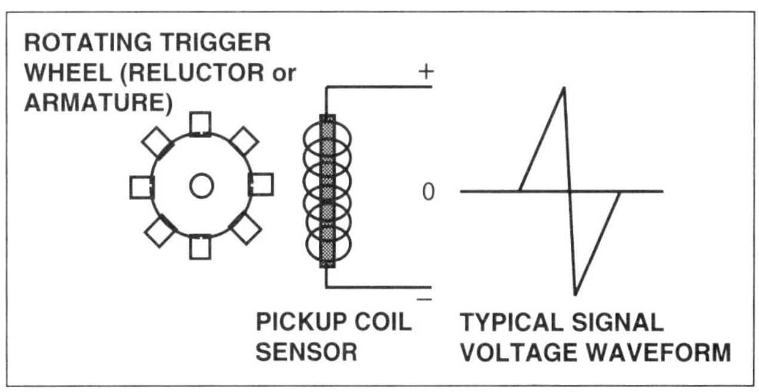

Figure 3-13. Resistance and signal voltage of a magnetic pickup sensor can be checked with a DVOM, and the signal waveform can be monitored with a lab scope.

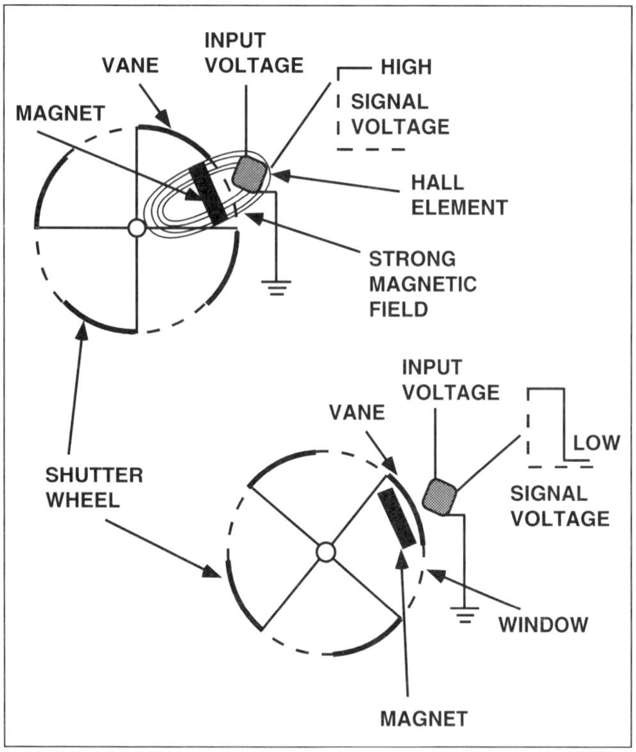

Figure 3-14. Hall-effect switches produce a variable frequency signal that is high when the magnetic field is strong, and low when the field is weak.

waveform as the signal toggles between high to low voltage. The frequency of the wave varies in proportion to distributor, or engine, speed.

Hall-effect switches seem to function better on a higher voltage signal than other electronic components, and are often an exception to the 5-volt reference rule. Keep this in mind when checking supply voltage, the unit may require 7, 9, or even 12 volts to operate properly. On engines that use more than one Hall-effect switch, such as crankshaft and camshaft sensors, a common circuit generally supplies the power.

A lab scope is the ideal tool for monitoring the signal wave produced by a Hall-effect switch. An alternative test method is to use a 12-volt battery, a voltmeter, and a steel feeler gauge or knife blade. Test as follows:

1. Connect the battery across the positive and negative source voltage terminals of the switch.
2. Connect the voltmeter across the signal voltage and negative terminals of the switch.
3. Insert the steel feeler gauge or knife blade between the magnet and the Hall-effect switch.
4. Observe the voltmeter. The reading should be within 0.5 volt of battery voltage. Remove the knife or feeler gauge and the reading should drop to less than 0.5 volt.

Oscilloscope Diagnosis

Because an oscilloscope displays voltage changes during a period of time, it is an ideal instrument for examining the variable voltages produced in both the primary and secondary ignition circuits. Oscilloscopes display a voltage trace on a viewing screen as a graph of voltage versus time. The vertical scale on the screen represents voltage and the horizontal scale represents time.

Both primary and secondary ignition system voltage traces can be divided into three sections: firing, intermediate, and dwell, figure 3-15. Deviations from a normal pattern indicate a problem. In addition, most scopes display ignition traces in three different patterns. You can use each pattern to isolate and identify particular kinds of malfunctions. The three basic patterns are:

- Superimposed
- Parade, or display
- Stacked, or raster.

In a **superimposed pattern**, voltage traces for all cylinders are displayed on top of each other to form a single pattern, figure 3-16. This display provides a quick overall view of ignition system operation and can also reveal certain major problems.

The **parade pattern** displays voltage traces for all cylinders one after the other from left to right across the screen in firing order sequence, figure 3-17. This allows easy comparison of voltage levels between cylinders. A parade display is useful for diagnosing problems in the secondary circuit.

A **raster pattern** shows the voltage traces for all cylinders stacked one above the other in firing order sequence, figure 3-18. This display allows you to compare the time periods of the three sections of a voltage trace.

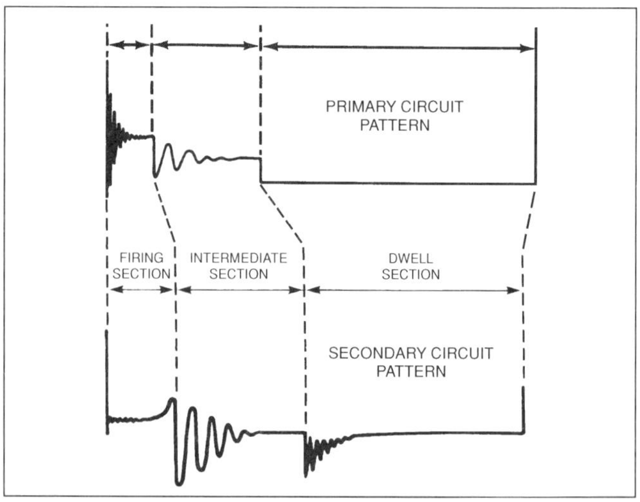

Figure 3-15. Ignition oscilloscope patterns, both primary and secondary, can be divided into three distinct sections.

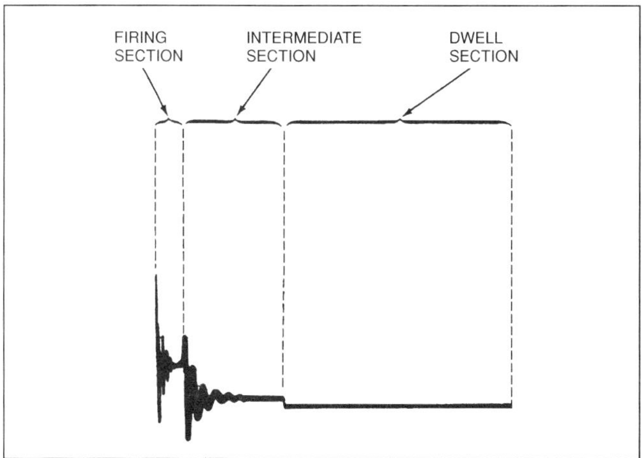

Figure 3-16. A superimposed pattern displays cylinder traces on top of each other.

Primary Oscilloscope Patterns

Oscilloscopes display primary scope patterns as a low-voltage trace moving from left to right across the screen. Electronic ignition patterns vary by system, so you must know what is normal for the system being tested. Examples presented here are intended as a general reference only.

Firing section

The firing section of the pattern corresponds to the amount of time the spark plug is firing. The voltage trace records induced voltage surges in the primary

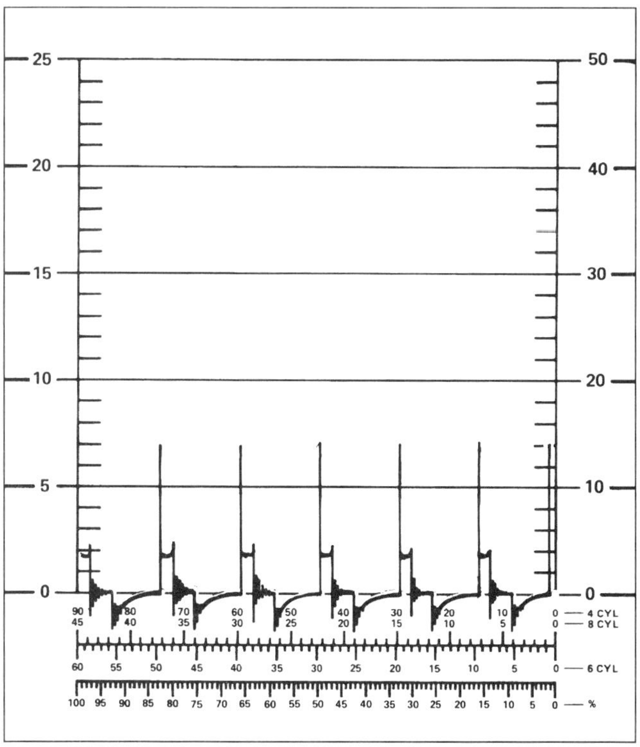

Figure 3-17. Parade pattern displays cylinder traces horizontally across the screen.

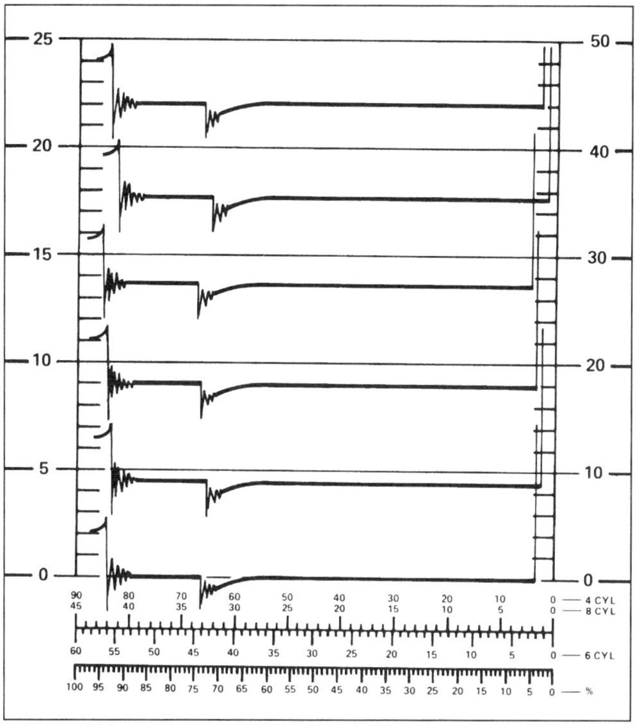

Figure 3-18. Raster pattern stacks cylinder traces vertically on the screen.

circuit as secondary voltage is being dissipated from the coil. A typical pattern shows a vertical rise as spark is established followed by a series of diminishing oscillations until spark is extinguished, figure 3-19. On some systems, there are no oscillations in the firing section. A normal display for these systems appears as a vertical spike as spark begins followed by a relatively flat line until spark ends.

In parade display, look for variations in the height of the oscillations between cylinders. High or excessive primary circuit resistance will cause irregularities in the firing section of the pattern, figure 3-20. The cause will

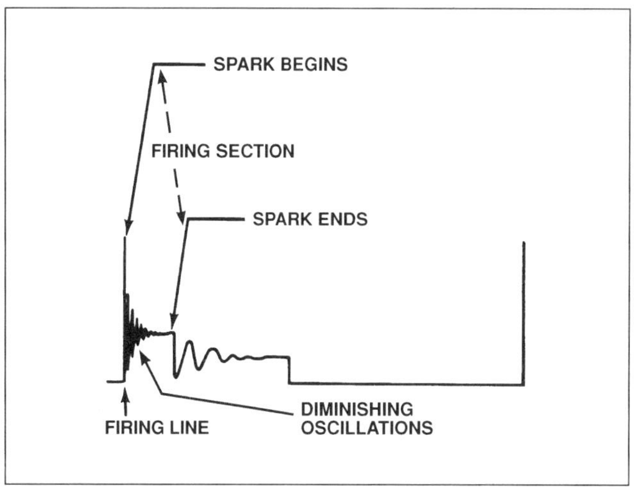

Figure 3-19. A normal primary firing section will have a high spike followed by a series of diminishing oscillations.

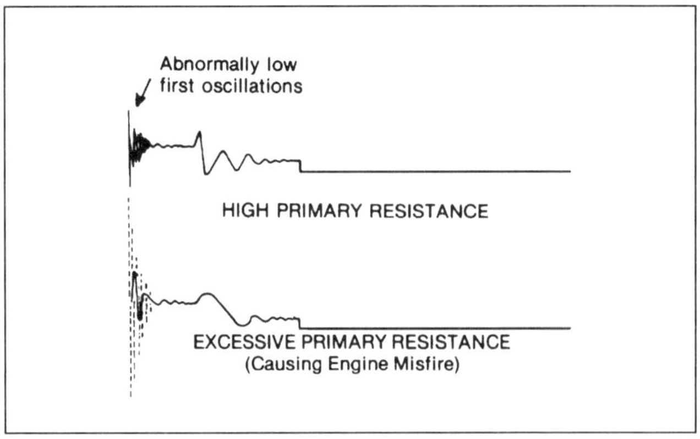

Figure 3-20. High, or excessive primary circuit resistance will disrupt the firing section trace.

generally be poor wiring connections or a defective coil. Excessive primary resistance can drop secondary voltage low enough to create a misfire that leads to increased exhaust emissions.

Intermediate section

Once spark ends, the intermediate section begins. This part of the trace continues until an electronic signal from the ICM switches on the primary circuit transistor. On some systems, such as the Chrysler EIS, there is no intermediate section because the primary transistor switches on the moment spark ends, figure 3-21.

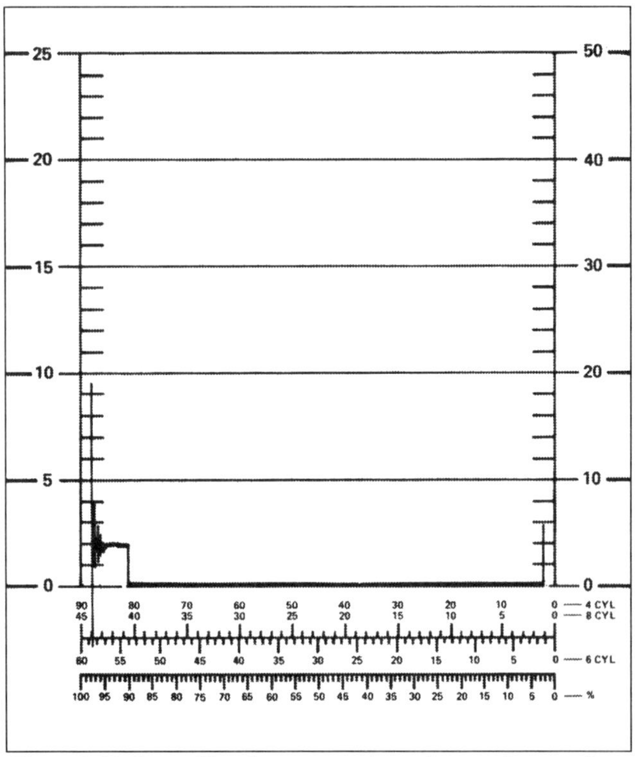

Figure 3-21. Some Chrysler systems do not have an intermediate section, the primary transistor switches on as soon as the spark extinguishes.

An intermediate voltage trace displays the dissipation of energy remaining in the coil after plug firing is complete. The trace appears as a series of diminishing oscillations that are similar to those of the firing section, but considerably smaller. There should be at least five distinct coil oscillations on the intermediate trace, figure 3-22. Less than five oscillations indicates a defective coil primary winding. The end of the intermediate section is marked by an abrupt drop. This is the point where the primary transistor turns on and dwell begins.

Dwell section

The dwell section is the time the primary circuit is complete. During dwell, a low-voltage current flow is building up the magnetic field of the coil. Following the sharp drop of the transistor switching on, the trace appears as a relatively flat horizontal line. The dwell section ends with the abrupt upward stroke of the first firing oscillation. Small oscillations, or a hump, towards the end of the dwell period are normal on some systems. If the ICM contains a current limiting device, the hump indicates the limiting circuit has been activated, figure 3-23.

Use the raster pattern to compare the dwell period for each cylinder. Dwell sections should not vary by more than four to six degrees between cylinders. Variations can be caused by a worn distributor or timing chain.

Secondary Ignition Patterns

Secondary voltage traces also move left to right across the screen in firing, intermediate, and dwell sequence, figure 3-15. However, since normal firing voltage can exceed 40,000 volts, oscillations are displayed on a high-voltage scale in kilovolt (kV), or thousand volt increments.

Firing section

The firing section of the secondary pattern starts with a straight vertical rise called the firing line or voltage spike. The peak of the spike indicates the amount of voltage required to create an arc across the spark plug air gap. Once an arc is established, it requires

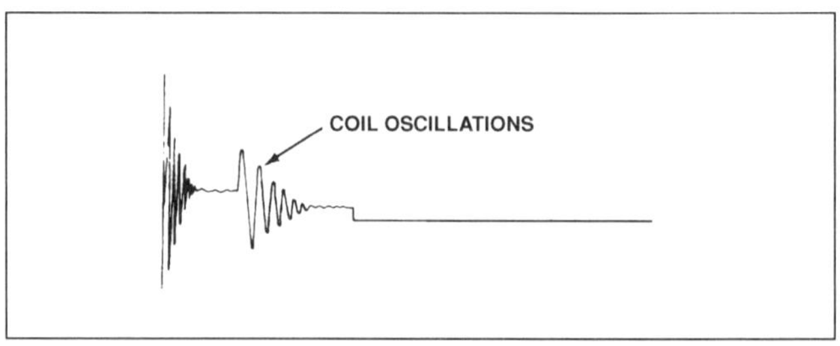

Figure 3-22. The coil is dissipating residual voltage during the intermediate section, there should be at least five distinct, diminishing oscillations.

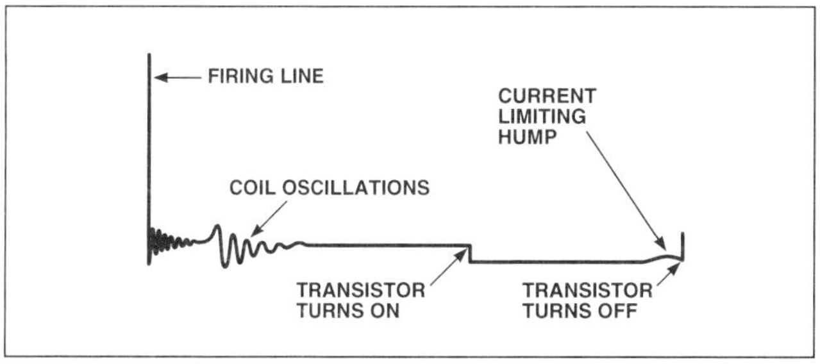

Figure 3-23. When a current limiting transistor is used, a small hump will appear on the trace at the end of the dwell section.

less voltage to maintain. A normal trace will drop to about one-quarter the height of the voltage spike, then continue as a horizontal spark line. A spark line represents continued current across the spark plug gap and should be relatively flat and stable. However, a series of very small oscillations in the spark line is not abnormal.

The firing section of a secondary trace can reveal a number of ignition problems. Damaged plug wires, worn distributor caps, and excessive plug gaps create high open circuit resistance. High resistance will generally appear as a high, short spark line, figure 3-24. Corrosion on the cable terminals also creates high resistance, but corrosion causes the spark line to start high and slope downward sharply. Carbon tracking in the distributor cap, poor cable insulation, and fouled spark plugs are sources of low resistance. Low resistance shows up as a low, long spark line, figure 3-25. A spark line that jumps erratically or slopes upward at the end indicates fuel mixture problems, figure 3-26. Mechanical problems, such as vacuum leaks or a burnt valve, can create a lean condition that can cause a similar upward sloping pattern.

Intermediate section

A secondary intermediate trace indicates excess coil voltage being dissipated, and will appear similar to a primary intermediate pattern. Look for a short vertical rise from the spark line followed by a series of diminishing oscillations. Oscillations should be of relatively even width and taper down gradually to a near horizontal line. The intermediate section ends with the module-on signal.

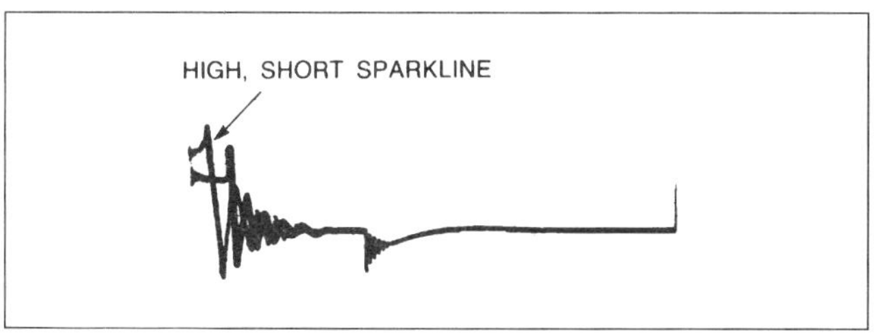

Figure 3-24. High open circuit resistance will produce a high, short spark line on the secondary trace.

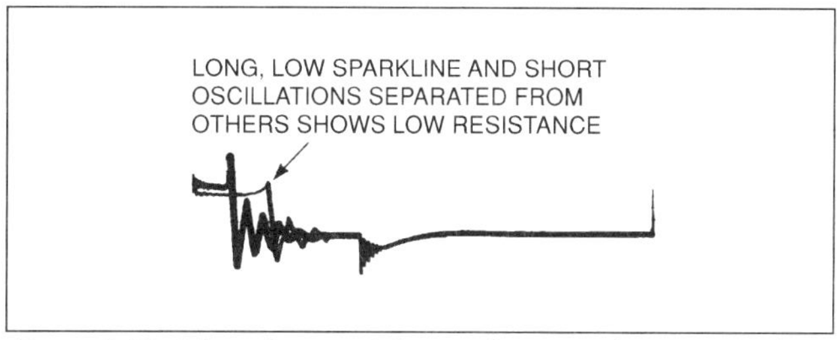

Figure 3-25. A long, low spark line on the secondary trace indicates low open circuit resistance.

A trace pattern that appears upside-down indicates the coil polarity is reversed, figure 3-27. An intermittent open in the coil secondary winding will cause the pattern to jump up and down on the screen, figure 3-28. Reduced oscillations without a module-on signal result when a problem exists between the coil and distributor cap.

Dwell section

The dwell section begins with the module-on signal and continues until the firing section for the next cylinder begins. Some module-on signals produce a series of oscillations, others do not. All will begin with a sharp vertical drop and continue as a nearly flat horizontal line. When oscillations are present, they should gradually diminish in size before leveling off to a near flat line.

The length of the dwell section will vary for different systems, and may also be engine speed dependent, figure 3-29. A slight rise, or hump, towards the end of the dwell period is normal if the ICM has a current limiting device.

ELECTRICAL NOISE

Electrical noise is a loosely defined term used to describe abnormal voltage levels on one circuit that result from electrical activity on a second circuit. The two

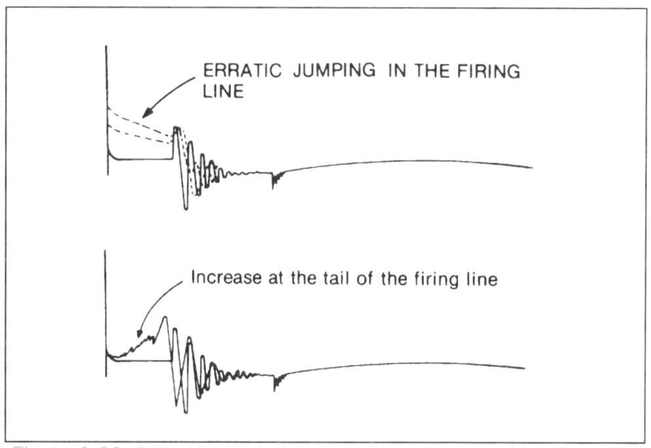

Figure 3-26. A spark line that jumps or slopes upward indicates fuel mixture problems. Mechanical failures, such vacuum leaks, bad fuel injectors, and burnt or sticking valves, are often at fault.

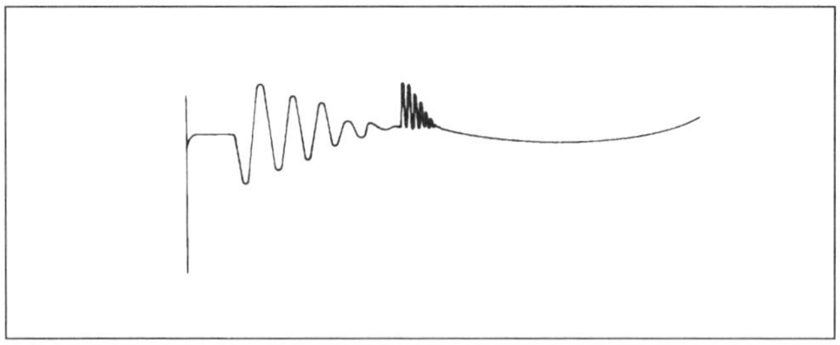

Figure 3-27. Reversed coil polarity will produce an inverted, or upside-down, trace.

circuits share no common connection. The interference, or noise, is induced through the insulation of the wiring.

An electromagnetic field is created whenever current flows through a conductor, the strength of the field changes when the current flow stops and starts. Each change in field strength creates an electromagnetic signal wave. The frequency of the wave increases in proportion to the speed of the current changes. When the signal wave is strong enough, it can be induced into an adjacent circuit.

In this phenomenon, known as electromagnetic interference (EMI), the signal wave is generated by the source and transferred into the receiver. EMI can be transmitted in four ways:

- Conductive coupling
- Capacitive coupling
- Inductive coupling
- Electromagnetic radiation.

A conductive couple occurs when interference is directly transmitted from the source to the receiver, figure 3-30. Capacitive coupling takes place when applied voltage creates an electrostatic field between the source and receiver, figure 3-31. An inductive couple forms when applied voltage creates an electromagnetic field between the source and receiver, figure 3-32. Electromagnetic radiation occurs when the receiver acts as an antenna and

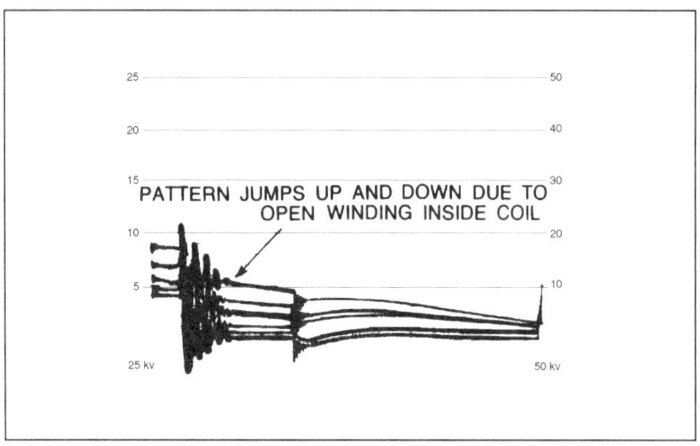

Figure 3-28. If the entire trace moves up and down on the screen, suspect an open in the secondary winding of the coil.

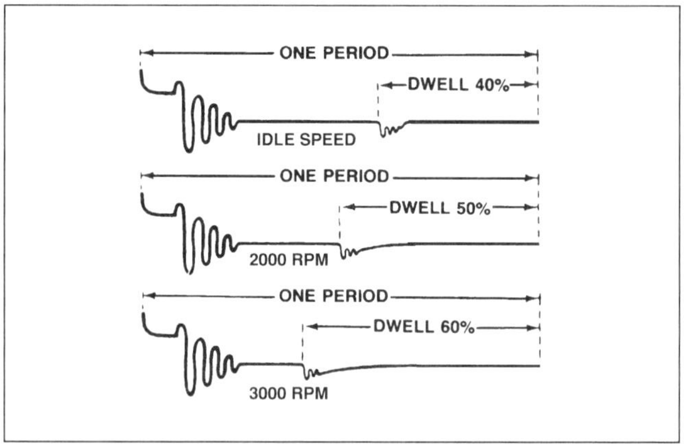

Figure 3-29. A dwell period that increases with engine speed is normal for some systems.

picks up signal waves generated by the source, figure 3-33.

In automotive applications, electrical noise can be classified into two categories: radio frequency interference (RFI) and electromagnetic interference (EMI). Although RFI is a form of EMI, radio frequency signals are much stronger and should be considered separately.

Radio Frequency Interference

Radio frequency interference produces a signal wave strong enough to interfere with radio and television broadcast transmissions. Secondary ignition circuits are capable of producing waves at the rate of thousands of cycles per second, or kilohertz (kHz). Signals of this magnitude fall into the range of commercial broadcast signals. RFI transmissions can also create serious problems on the low-power circuits of an automotive computer control system.

To prevent RFI that can effect commercial broadcasts, automakers are required to use television-radio-suppression (TVRS) cables on the secondary ignition system. These cables have a high-resistance non-metallic conductor core, such as carbon, linen, or graphite impregnated fiberglass, wrapped in several

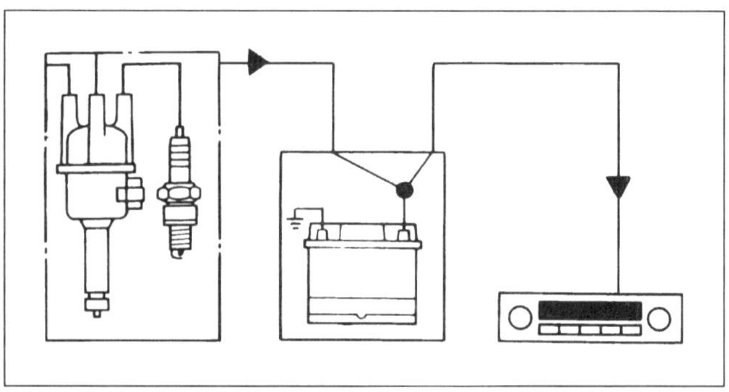

Figure 3-30. A conductive couple forms when a signal from the source is transmitted directly to the receiver.

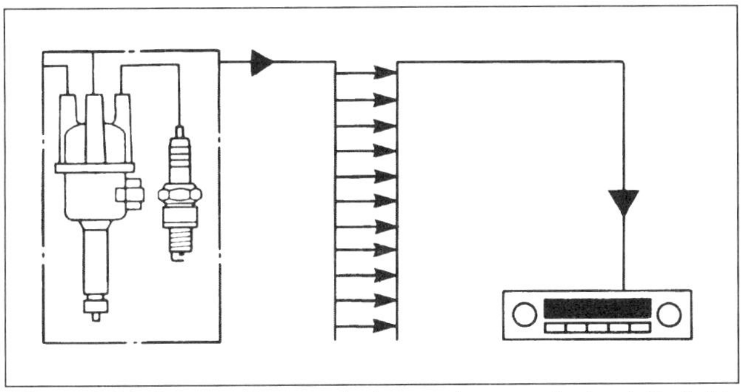

Figure 3-31. A capacitive couple forms when an electrostatic field forms between the source and receiver.

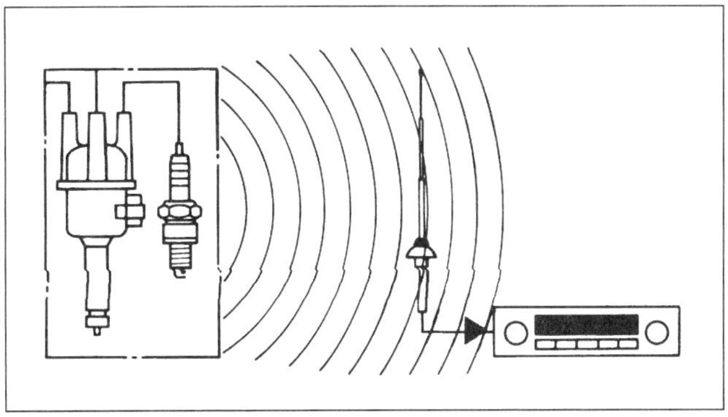

Figure 3-33. Electromagnetic radiation is when the receiver acts as an antenna to pick up signals generated by the source.

layers of insulating material, figure 3-34. Due to their high resistance, TVRS cables raise the voltage applied to the spark plugs. This offers the additional benefits of improved ignition characteristics and prevents the formation of some plug deposits. Higher voltage at the spark plug permits the combustion of leaner air-fuel mixtures, and also reduces the amount of current present after ignition is complete. This reduction of current can extend the life of the spark plugs and the distributor cap terminals.

RFI transmission generally is of little concern because vehicle manufacturers have taken steps to eliminate the problem. However, replacement ignition cables that do not meet original equipment specifications, or are incorrectly installed, can cause unwanted signals to be induced into other circuits. The resistance of secondary ignition cables will vary by length. A good TVRS cable will test at about 4,000 ohms per foot on a ohmmeter.

A common RFI problem is cross firing. Cross fire occurs when voltage from one cable is induced into an adjacent cable causing a spark plug to fire out of sequence. To prevent cross fire, route cables so they do not touch, and avoid running them parallel to each other. This parallel rule also applies to any primary wiring and harness that are in close proximity to the ignition cables. Routing the conductors so they cross each other at an angle disrupts the magnetic field and prevents induced voltage.

Electro-magnetic Interference

The revolving commutator in motors, the natural fluctuations in the alternator output circuit, electromechanical voltage regulators, and vibrating horn contacts all produce the transmission of electro-magnetic interference (EMI). In addition, whenever a relay or switch opens or closes a circuit or when a solenoid is activated, a wave is created. Static electric charges that develop from the friction of tires contacting the road, drive belts turning pulleys, axles and drive shafts rotating, and clutch and brake applications can also produce EMI transmissions.

Manufacturers are aware of this problem and take steps to suppress EMI on suspect circuits. Common suppression methods include the use of capacitors, suppression filters, ground straps, and shields.

A capacitor is a device that absorbs and stores voltage. When installed at the switching point of a circuit, a capacitor absorbs changing voltage and prevents EMI transmission. Manufacturers often install capacitors on the primary circuit at the ICM, figure 3-35, and on the output terminal of the alternator, figure 3-36. Suppression filters also absorb unwanted voltage

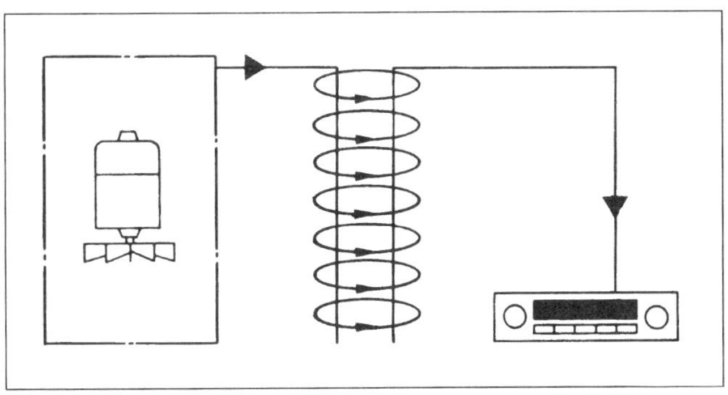

Figure 3-32. Inductive coupling occurs as a magnetic field forms and collapses between the source and receiver.

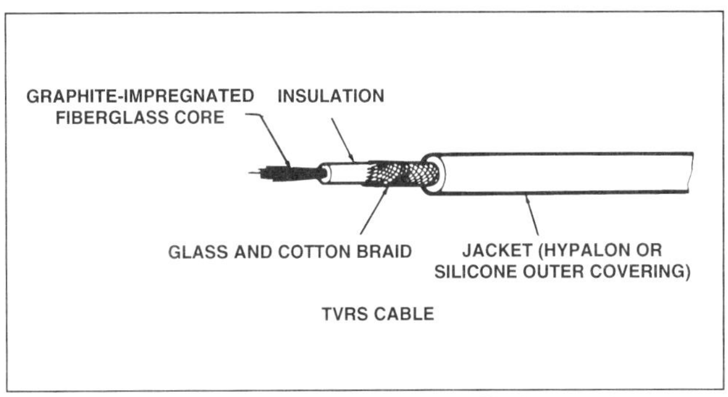

Figure 3-34. TVRS cables have a non-metallic, high-resistance core surrounded by several layers of insulation.

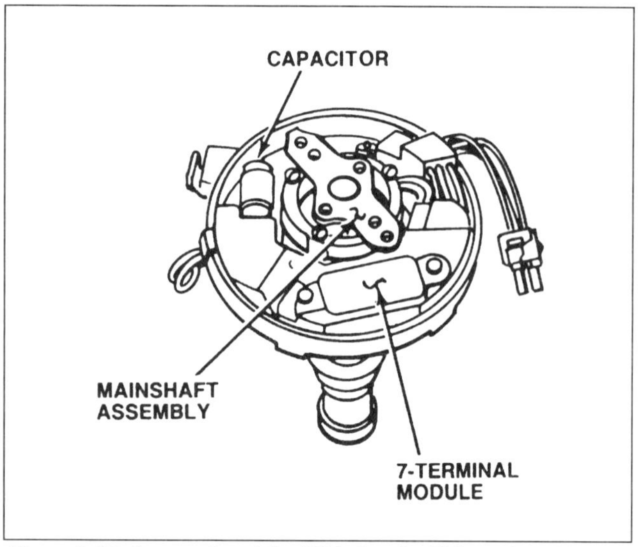

Figure 3-35. A capacitor at the ICM absorbs EMI produced by the primary circuit switching on and off.

signals. Filters and are often used to suppress the natural voltage fluctuations produced by the commutator of an electric motor, figure 3-37. Most replacement parts do not include the capacitor or suppression filter. It is up to you to transfer any capacitive suppression device from the old part to the new.

Ground straps provide an unrestricted, low-resistance circuit ground path to help suppress EMI conduction and radiation. In a ground path, resistance creates an unwanted voltage drop. The voltage difference across the resistance can be an EMI source. Ground straps are typically installed between body parts and bushing-mounted components to ensure good conductivity. Some late-model cars have a ground strap connecting the hood to the fender or firewall. These straps prevent an electro-static field from forming between body panels that could couple with harnesses mounted to the sheet metal.

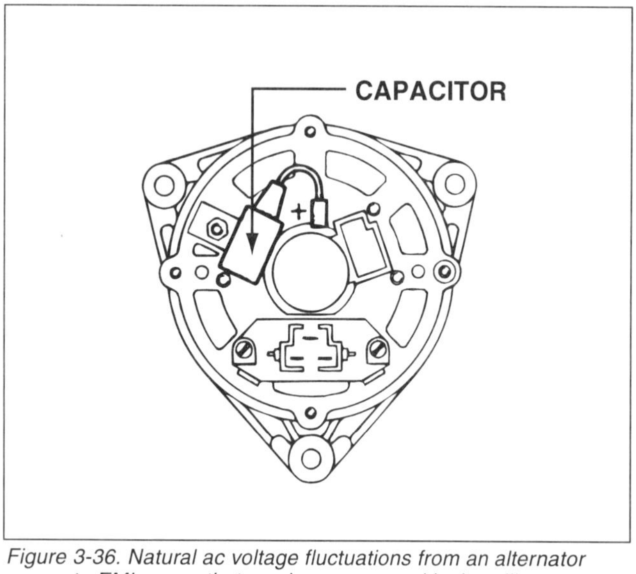

Figure 3-36. Natural ac voltage fluctuations from an alternator generate EMI waves that can be suppressed by installing a capacitor on the output terminal.

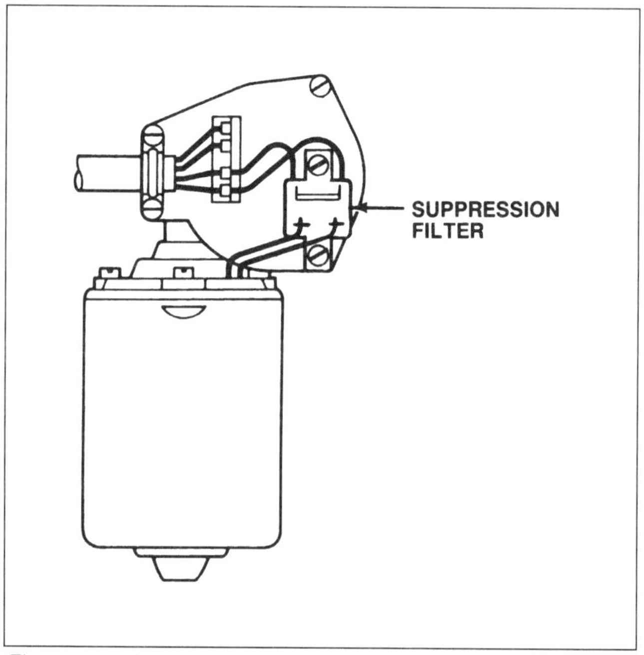

Figure 3-37. Suppression filters absorb interference generated by electric motors.

A simple metal shield surrounding a component capable of generating EMI signals will effectively block any wave transmission. Several manufacturers also use a metal shield on their distributors to control RFI radiation. In addition, the metal housing of an on-board computer acts as a shield to prevent EMI waves from entering the unit and causing internal damage.

Any time you remove a suppression device when servicing a vehicle, it is important to reinstall it on assembly. Problems related to EMI transmissions are difficult to diagnose because they can be random and unpredictable. Use a DVOM, scan tool, or lab scope to monitor suspect circuits. Then, watch for abnormal circuit activity as you activate switches, relays, solenoids, and motors. Also make sure your body is grounded when you handle components that contain a microprocessor. Simply sliding across a velour seat can generate enough static electricity to damage delicate computer circuits.

1. A customer brings his car into your shop for a tune up. The car runs rough and lacks power. During a cylinder power balance test you discover cylinder #1 is weak, and you see the secondary scope pattern shown.

   Technician A says this could be caused by a burnt valve.

   Technician B says this could be caused by a fouled spark plug.

   Who is right?

   a. A only
   b. B only
   c. Both A and B
   d. Neither A nor B

   Explanation: Both fuel and compression affect the firing line. Since fuel conducts electricity it takes more voltage to jump the gap if the engine is lean. The burnt valve resulted in a loss of compression and a lean fuel condition. At the end of the firing duration on cylinder #1, as the mixture leaned out, the voltage requirements went up, causing the voltage of the firing line to rise. Technician A is correct.

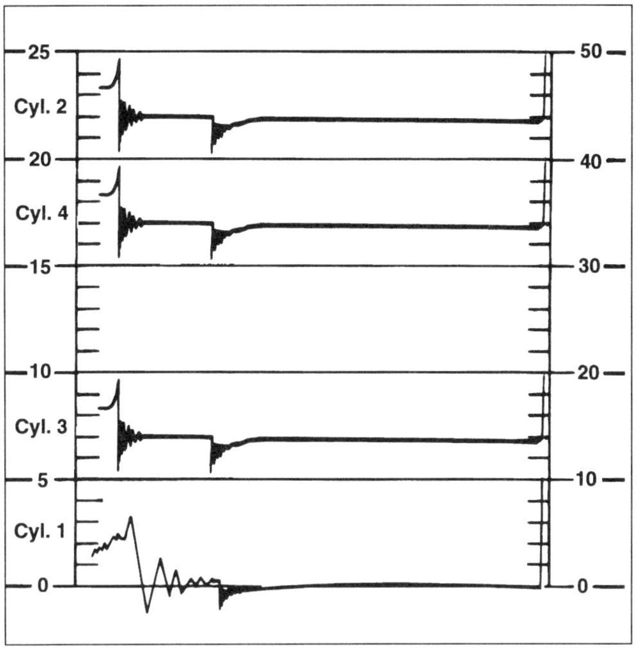

Illustration for question 1.

2. On the composite vehicle, pins #24 ECT and #25 IAT are both shorted to ground.

   Technician A says this will cause the MAP and TP sensor voltage to go low.

   Technician B says this will cause the 5 volt reference at pin #13 to go low.

   Who is right?

   a. A only
   b. B only
   c. Both A and B
   d. Neither A nor B

   Explanation: On the composite vehicle, pins #24 ECT and #25 IAT both individually receive there own 5 volt reference from the PCM. Therefore, if they short to ground they will not have any affect on the voltage reference to the MAP or TP sensors. Neither Technician is right.

3. While replacing the valve cover gaskets on the composite vehicle, a technician inadvertently pinched pin #13 causing the wire to short to ground.

   a. This will cause the MAP sensor and TP sensor voltage to go low.
   b. This will cause the MAP sensor, TP sensor, ECT sensor, and IAT sensor voltages to go low.
   c. This will cause the MAP sensor, TP sensor, ECT sensor, IAT sensor and O2 sensor heater voltages to go low.
   d. This will cause the MAP sensor, TP sensor, ECT sensor, IAT sensor, IAC motor and O2 sensor heater voltages to go low.

   Explanation: When pin #13 is shorted to ground only the MAP sensor and TP sensor receive the 5 volt reference from pin #13. If you look closely at the composite vehicle's wiring diagram, figure 3-1, the other sensors receive their own voltage reference and only use the ground side of the shorted signal. So the short will not effect these signals on the composite vehicle. Technician A is correct.

4. Technician A says that open circuit battery voltage should be 12.4 volts for proper cranking. If not, charge the battery before you begin electronic diagnostics.

   Technician B says most systems require at least 10 volts to operate, and if voltage is low the system may not be able to complete a diagnostic sequence.

   Who is right?

   a. A only
   b. B only
   c. Both A and B
   d. Neither A nor B

   Explanation: Open circuit battery voltage below 12.4 volts may not be sufficient to properly crank the starter motor. Although most computer-controlled circuits operate on low voltage, the PCM must be supplied with system voltage in order to operate. When system voltage drops below 10 volts, the PCM may store a code and suspend operations. Therefore, both technicians are correct.

5. Where would you look to find base timing information for the vehicle you are troubleshooting?
   a. The owner's manual
   b. The SEFI decal
   c. The Fast Track Troubleshooter
   d. The emissions label

   Explanation: The emissions label will normally provide basic timing, idle speed and emissions control information for the vehicle, so d is the correct answer.

6. Technician A says that having several ground points in a complex electronic circuit spreads the voltage drop around so it cannot cause control system problems.

   Technician B says that small voltage offsets between ground points may cause control system problems.

   Who is right?
   a. A only
   b. B only
   c. Both A and B
   d. Neither A nor B

   Explanation: Statement A points out a misunderstanding of the term "voltage drop". A voltage drop occurs in any circuit wherever there is resistance. If there is *extra* resistance in a circuit, there may be an *excessive* voltage drop. Technician B is correct. Small voltage offsets anywhere in a control system circuit may cause problems because these are generally low-voltage circuits.

7. Technician A says high resistance in a circuit can cause system problems because control systems use low voltage components.

   Technician B says low resistance in a component can cause system problems because it may cause a high current condition.

   Who is right?
   a. A only
   b. B only
   c. Both A and B
   d. Neither A nor B

   Explanation: Both statements are correct and point out the need to check component specifications carefully if there is a problem. High resistance in a circuit can cause a relatively large voltage drop in a low voltage circuit. Low resistance in a component can cause increased current flow which can damage components.

8. The composite vehicle's engine is at idle. What voltage should a technician expect to see at pin #41?
   a. 0.0 volts
   b. 0.05 to 0.75 volts
   c. 4.5 to 5.5 volts
   d. 12 to 14 volts

   Explanation: Anytime the fuel pump relay is energized, the PCM is pulling the voltage low at pin #41. If the voltage was 0, then there would be no voltage to the relay. If the voltage was near battery voltage, then the relay would not be energized. The answer is B; a voltage between approximately 0.05 to 0.75 indicates a normally energized fuel pump.

Chapter Four

FUEL SYSTEMS AND AIR INDUCTION SYSTEMS DIAGNOSIS

An engine depends upon a constant mixture of fuel and air blended at, or near, the stoichiometric ratio for efficient operation. In addition, the engine must be mechanically sound. Any leakage, whether fuel, air, or compression, has an effect on the air-fuel ratio. This leads to inefficient combustion and can result in a loss of performance and an increase in emissions. This chapter covers diagnosing engine efficiency problems and recognizing symptom failures that relate to the air intake and fuel delivery systems. We begin with a brief discussion of the L1 composite vehicle fuel injection system, and continue with diagnostic procedures and component checks.

COMPOSITE VEHICLE FUEL SYSTEM

The sequential multiport fuel injection (SFI) system of the composite vehicle is similar to many of the electronically controlled fuel systems in actual use. In the following paragraphs, we discuss the operation of the composite vehicle system, as well as provide basic information on some of the variations you are likely to encounter in the field.

System Operating Modes

Fuel injection strategy can be broken down into three modes of operation:

- Staring mode
- Clear flood mode
- Run mode.

Starting mode

With the ignition switch on, battery voltage is available at pin 11 of the PCM. In response, the PCM momentarily energizes the fuel pump relay by completing the pin 12 circuit. This directs battery voltage to the fuel pump and pressurizes the system. To prepare for start-up, the PCM monitors coolant and air temperature, atmospheric pressure, and throttle position inputs to determine initial injector pulse width. As a fail-safe, the PCM opens the relay to shut off the fuel pump if the ignition is left on for more than 2 seconds without cranking the engine. Once the engine cranking begins, the fuel pump relay remains energized as long as engine speed exceeds 100 rpm.

Clear flood mode

To facilitate starting a flooded engine, the PCM program switches off the fuel injectors when the starter is engaged, the throttle position signal indicates a wide open throttle (over 80% open), and engine speed is below 400 rpm. Any problem that causes the throttle signal to go high (over 3.7 volts) can place the system in clear flood mode and result in a no-start condition.

Run mode

Run mode includes the open and closed loop operation, acceleration enrichment, and deceleration enleanment strategies. A discussion of basic loop operating strategies can be found in Chapter 1 of this book, and a summary of composite vehicle operation appears as figure 1-12.

Fuel cut-off mode

As a safety precaution on the composite vehicle, the PCM program also contains a fuel cut-off mode. Should vehicle speed reach 110 mph, or engine speed exceed 6000 rpm, the PCM switches off the fuel injectors to prevent engine damage. Systems on other vehicles may switch off ignition, instead of fuel, as the safety precaution. The speed or rpm at which cut off occurs may be different, depending on the system.

Fuel Delivery and Management

An electric pump, powered by a PCM controlled relay, pressurizes the fuel system. Manifold vacuum controls a mechanical pressure regulator to maintain a constant 34 to 38 psi (234 to 262 kPa) of fuel pressure across the injectors.

Variable pulse-width injectors fire individually, once per camshaft revolution. Injectors fire in response to input signals from the engine speed sensor. A feedback signal from the oxygen sensor, as well as input signals from the throttle position sensor, coolant and intake air temperature sensors, manifold absolute pressure sensor, and engine speed sensor are used to calculate injector pulse width, or on time. A diagram of composite vehicle components appears as figure 4-1. Also, see figure 3-1 in the previous chapter for an electrical system diagram.

With sequential injection, a dedicated driver circuit is used to control each individual injector. Common alternatives are to fire the injectors in pairs with one driver controlling two injectors, and in banks, so one driver would fire three injectors on a V6 engine. On these systems, one failed injector can cause operating problems and inaccurate test results from its paired counterparts. Knowing how the injectors are driven is important when organizing your approach to isolating faulty components.

Intake Air Management

Electronic control systems maintain correct air-fuel ratio by adjusting injector pulse width based on the amount of air entering the engine. To determine the correct amount of fuel, the PCM must be able to accurately measure the intake air charge. The PCM of the composite vehicle calculates the amount of intake air by the speed density method. Two other designs, mass airflow and volume airflow, are also common. Each are discussed below:

Speed density

The PCM of a speed density system estimates the density of the intake air charge to regulate fuel metering. Air density estimates are internal PCM calculations based on the input of various sensors. To estimate, the PCM accesses lookup tables using manifold absolute pressure (MAP) and engine speed signals as the primary inputs. Signals from the air and coolant temperature sensors, and other inputs, are used to fine tune the fuel mixture.

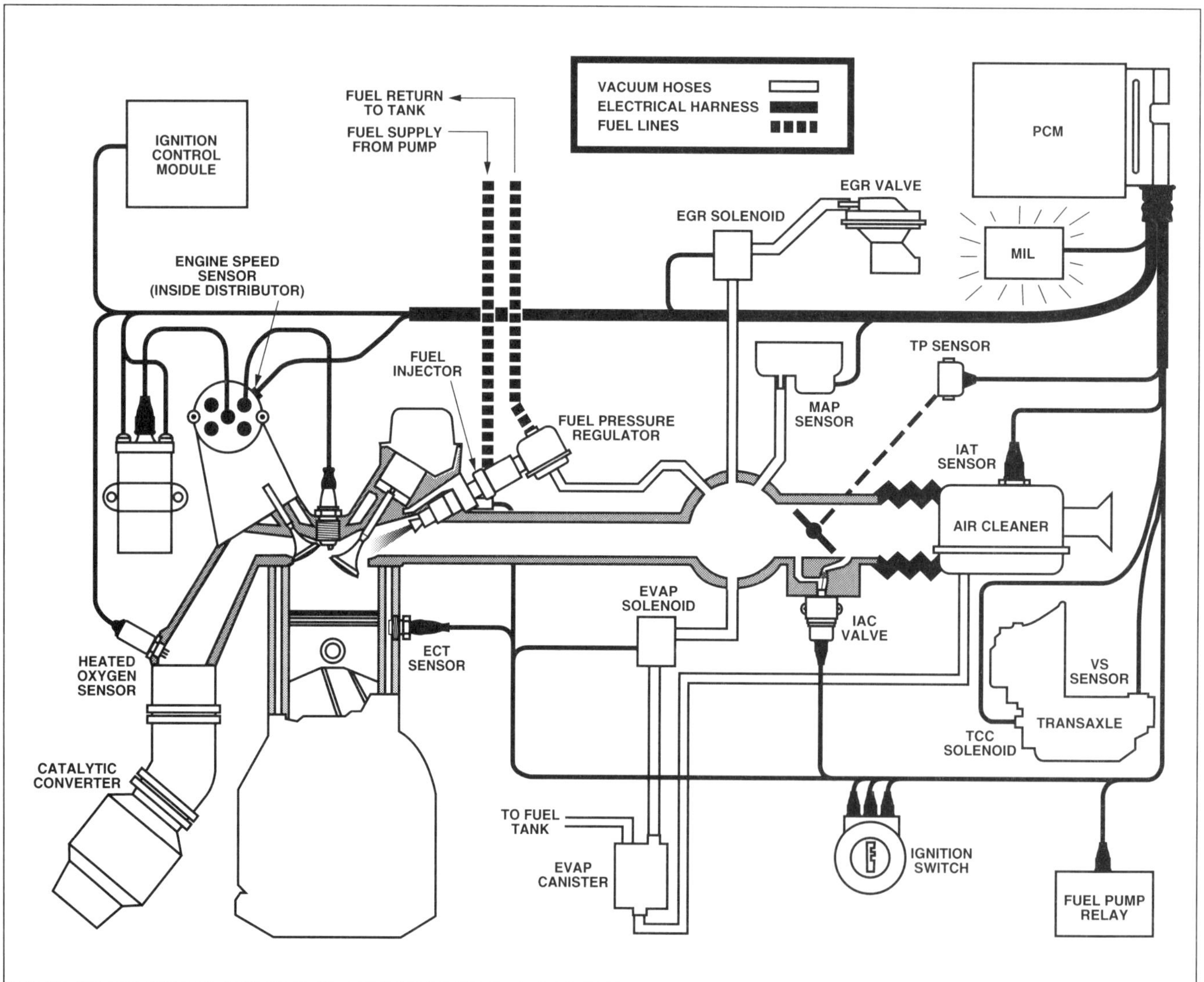

Figure 4-1 Composite vehicle engine management system.

Failure on any of the input circuits can result in a false estimate of air density, which results in an incorrect air-fuel ratio. This can cause driveability problems and emission test failure. Most systems will enter a back-up, or limp-home, program in the event of a severe input failure. Be aware, back-up parameters may be all that is available on the serial data stream. Electronic diagnosis is detailed in Chapter 2.

Mass airflow

The PCM of a mass airflow (MAF) system also estimates the density of the intake air charge to regulate fuel metering. However, the calculation is based on the amount of electrical energy required to maintain a constant temperature across a heater element or heated wire placed in the intake air stream. Both increases in air flow and density create a demand for electrical current as they draw more heat off of the element. As explained in Chapter 3, the heater element is part of the MAF sensor which transmits an input signal that increases in frequency along with mass airflow. This type of airflow meter is used by General Motors. The Bosch hot wire mass airflow sensor uses a heated wire to output an analog voltage signal that increases with airflow.

Volume airflow

With a volume airflow system, the PCM input calculates fuel correction based on the volume of air entering the engine. A mechanical device, a movable vane placed in the air stream, measures air volume. A potentiometer attaches to the vane to provide a variable voltage input signal to the PCM.

Idle Speed Management

The composite vehicle controls idle speed with a stepper motor. The stepper motor regulates the amount of air allowed to bypass the throttle plates. The serial data stream displays stepper motor position as the idle air control (IAC) value. IAC value is a percentage of bypass opening, with 0% as no bypass air and 100% as full bypass air flow.

Base idle is not adjustable, and the system should maintain a no-load idle speed between 850 and 900 rpm. Under normal circumstances, the IAC value will range from 15% to 25% at idle.

DRIVEABILITY PROBLEM AND EMISSIONS FAILURE DIAGNOSIS

When a control system is unable to maintain the correct air-fuel ratio, driveability suffers. Typical complaints include hesitation and stumble, lack of performance, and poor fuel economy. All of these will effect tailpipe emissions, and all can be the result of a mechanical problem. Eliminate mechanical problems first, before concentrating on the fuel system.

Generally, you begin diagnosis by checking fuel pressure to eliminating fuel delivery problems. Often, you can verify a rich or lean condition by scanning injector pulse width. A high pulse width indicates more on-time and a rich mixture, and low pulse width indicates less on-time and a lean mixture. Be careful, as pulse width may be responding to a false feedback signal generated by the oxygen sensor. In addition, there are no established specifications for injector pulse width. However, readings should change in proportion to engine speed and load. Typical readings will range from about 1 to 4mS at idle to around 12mS at wide open throttle.

If system pressure checks out good, and injector pulse width is responding, suspect an airflow problem. Typical causes include incorrect signals from the MAP and TP sensors. Also look for throttle body and idle air control valve airflow restrictions. Keep in mind, any disruption of the intake air can result in an erratic input signal being transmitted to the PCM. Since the signal does not correspond to the PCM program, it cannot be processed. This makes it impossible for the PCM to precisely control fuel delivery, and often results in a driveability problem that does not record a diagnostic trouble code.

In general, three factor come into play:

- Engine mechanical condition
- Fuel delivery system
- Air induction system.

ENGINE MECHANICAL CONDITION

When an engine is running, the downward motion of the pistons on their intake strokes reduces pressure in the cylinder and the intake manifold to produce a vacuum. The total vacuum produced depends upon two factors:

- **Engine Design** — Number of cylinders, displacement, compression ratio, valve overlap and lift — all help govern the amount of vacuum an engine produces.
- **Engine Condition** — Piston ring and valve wear can cause a measurable vacuum loss, as can manifold leakage, fuel problems, incorrect ignition timing, and some exhaust failures.

Since a worn engine cannot produce as much vacuum as one in good condition, measuring vacuum is a good way to determine engine condition. However, engines with emission controls often have high lift cams with greater valve overlap, this results in a lower level of vacuum with a less steady reading than those without emission controls.

Vacuum Sources

Vacuum drawn directly from a tap at the intake manifold, figure 4-2, is called manifold vacuum. Manifold vacuum is used to diagnose engine condition. Manifold vacuum is greatest at idle and decreases until there is very little at wide-open throttle. On the composite vehicle, the fuel pressure regulator, the MAP sensor, the EGR valve and solenoid, and the canister purge solenoid all require manifold vacuum to operate.

Vacuum drawn from a tap just above the throttle plates is called ported vacuum. When the throttle is closed, there is no, or little vacuum at the port. As the throttle opens, airflow across the port increases to create a vacuum, figure 4-3. Ported vacuum cannot exceed manifold vacuum, and is often used to operate distributor advance units and some emission control components.

Engine Vacuum

A vacuum gauge is used to measure the difference between atmospheric pressure and intake manifold pressure, or vacuum. Normal manifold vacuum on an engine running at idle usually varies between 18 and 22 inches of mercury (in-Hg) (60 and 74 kPa) at sea level. Vacuum readings will drop about one inch or 25 millimeters for every 1000 foot (305 meter) increase in elevation above sea level, figure 4-4.

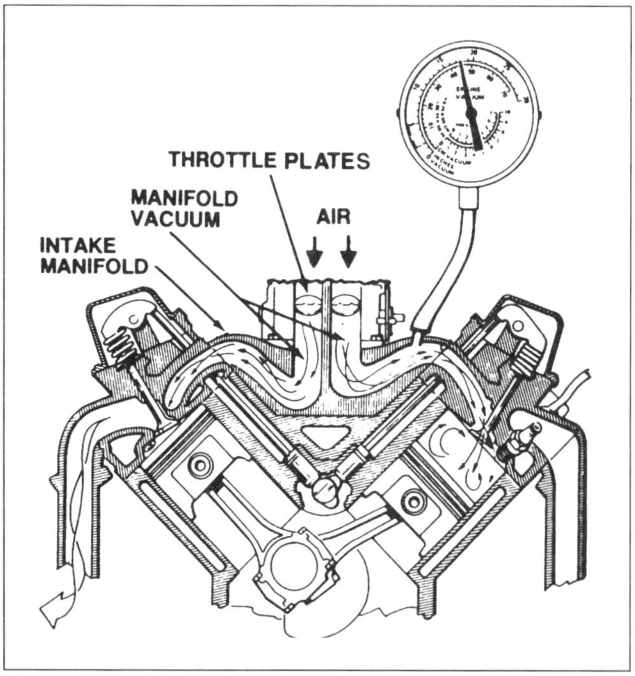

Figure 4-2 Manifold vacuum is the low pressure zone between the throttle plates and intake valves.

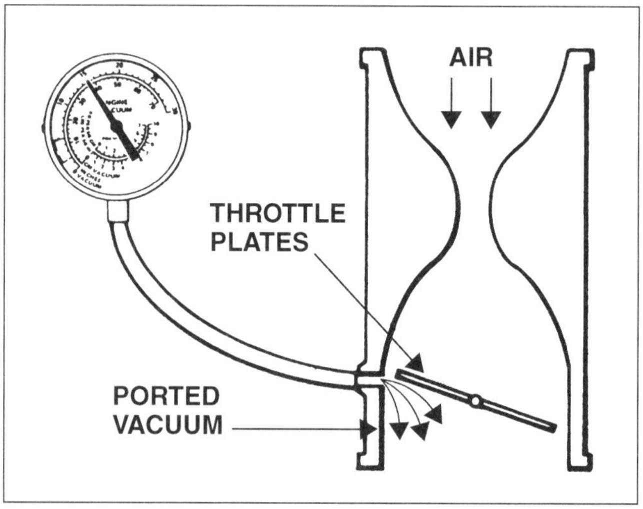

Figure 4-3 Ported vacuum is the low pressure created just above the throttle plates as they open.

Cranking Vacuum diagnosis

Measuring manifold vacuum while the engine is cranking is a valuable guide to determine engine compression and intake system conditions. High and constant gauge readings indicate the engine is mechanically sound and has a properly sealed intake manifold. Low vacuum at cranking speed can mean poor engine compression or leakage at the intake manifold.

To test, measure engine cranking speed with a tachometer and disable the ignition system to prevent startup. Crank the starter and note the vacuum gauge reading and engine speed.

Expect to see a steady vacuum reading of 5 in-Hg (17 kPa) or more with an engine in good condition. The needle may pulsate or wobble slightly, but it should be even and rhythmic. The cranking speed should be steady and normal. Readings of less than 5 in-Hg (17 kPa) indicate leakage. Perform a cylinder leakage or compression test to pinpoint the problem. If the readings are below 1 in-Hg (4 kPa), disconnect the throttle stop solenoid or the IAC motor to increase cranking vacuum and repeat the test.

PCV diagnosis

To check PCV flow, disable the ignition and pinch the PCV hose with a pair of pliers while cranking the en-

ALTITUDE	MANIFOLD VACUUM (in-Hg)
Sea Level to 1,000 Ft.	18 to 22
1,000 Ft. to 2,000 Ft.	17 to 21
2,000 Ft. to 3,000 Ft.	16 to 20
3,000 Ft. to 4,000 Ft.	15 to 19
4,000 Ft. to 5,000 Ft.	14 to 18
5,000 Ft. to 6,000 Ft.	13 to 17

Figure 4-4 Manifold vacuum varies by elevation.

gine. The vacuum gauge reading should increase if the PCV valve is open and the system is working properly.

Vacuum leakage diagnosis

Connect the vacuum gauge to a manifold vacuum source, connect a tachometer to the engine, and start the engine and run until normal operating temperature is reached. The vacuum gauge needle should hold a steady reading of about 15 to 21 in-Hg (51 to 71 kPa) with a warm engine running at idle. Abnormal gauge readings may be caused by:

- Valve seating problems
- Valve guide wear
- Head gasket leakage
- Intake manifold leakage.

Engine valve seating problems

Vacuum leaks caused by valve leakage can cause the needle to drop 1 in-Hg (4 kPa) or more below the normal reading. Each time a valve fails to seat, the needle fluctuates or flutters, figure 4-5. Locate leaking valves with a compression test or a cylinder power balance test.

Worn intake valve guides

Guide wear can also cause slightly below normal gauge readings. However, the needle will vibrate rapidly at idle and stabilize as engine speed increases, figure 4-6.

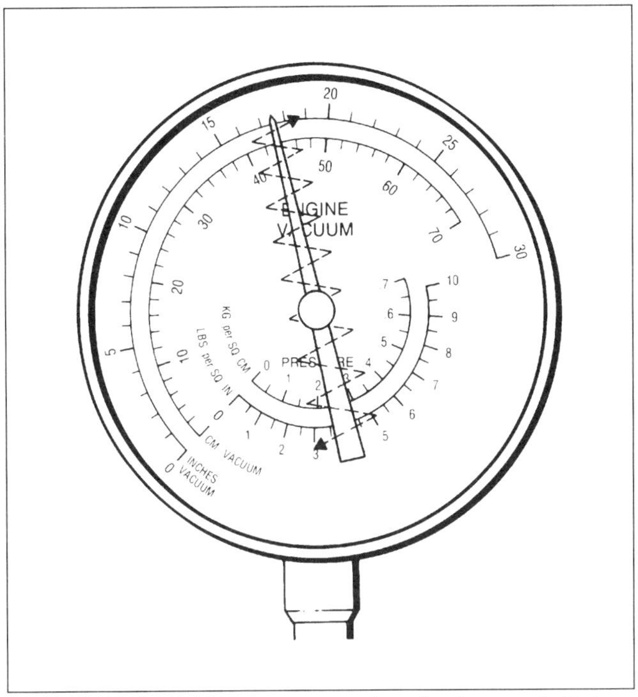

Figure 4-6 Valve guide wear causes a low vibrating needle that stabilizes as speed comes up.

Head gasket leakage

A leaking head gasket will cause the needle to vibrate as it floats through a range from below minimum to above maximum normal readings, figure 4-7.

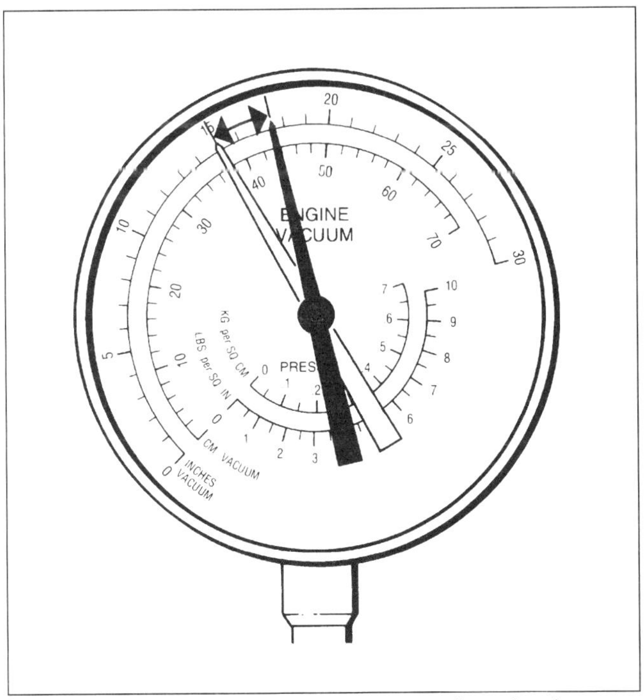

Figure 4-5 A valve not seating causes the needle to drop in a rhythmic pattern.

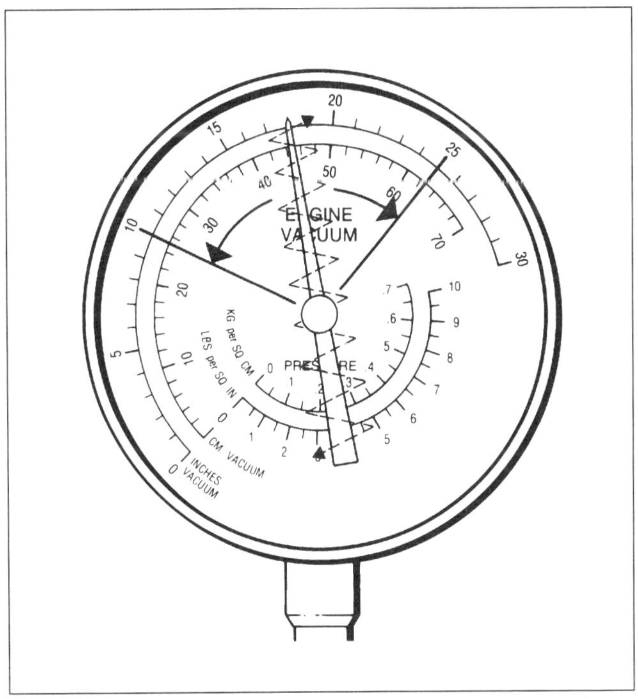

Figure 4-7 Head gasket leaks cause the needle to vibrate as it sweeps low to high.

Intake leakage

A gauge needle that fluctuates constantly from three to nine inches (75 to 230 mm) below normal indicates a vacuum leak somewhere else in the intake system, figure 4-8. Defective manifold gaskets and throttle body mounting gaskets are the most common sources of these leaks.

Valve action diagnosis

Correctly interpreting vacuum gauge readings and needle action can detect a variety of valve train problems, including:

- Valve timing
- Valve action
- Valve spring condition.

Valve timing

Excessively late valve timing on an idling engine will cause the needle to fluctuate between 5 and 7 in-Hg (17 and 24 kPa) on the gauge. Check for excessive slack in the timing belt or chain, and for incorrect timing component installation.

Valve action

An irregular drop from a normal reading indicates one or more sticking valves, figure 4-9. The action of the needle is similar to that of a leaking valve, but not as consistent.

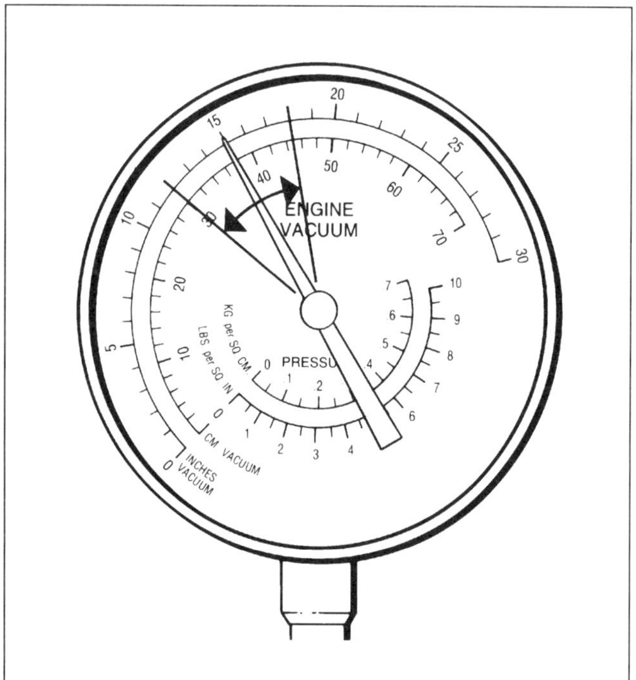

Figure 4-8 Intake air leaks cause a low fluctuating needle.

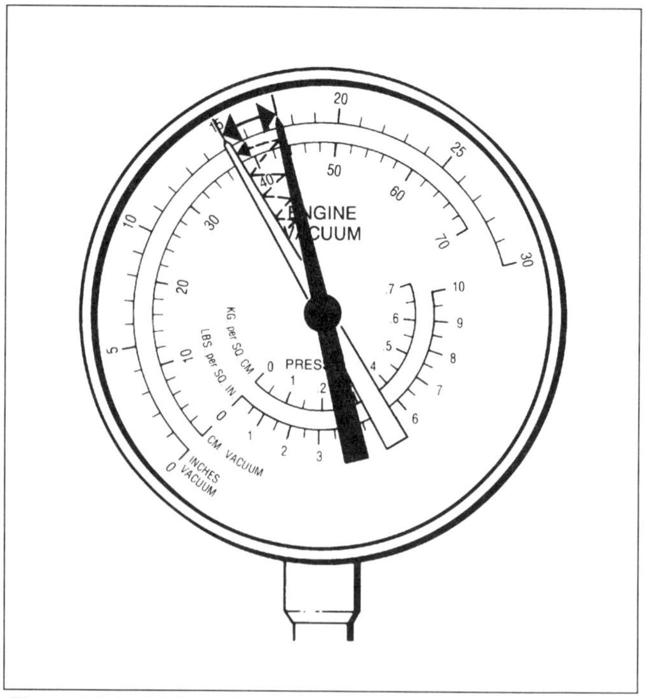

Figure 4-9 A sticking valve causes an irregular drop of the needle.

Valve spring problems

Increase engine speed to about 2,000 rpm to check for weak or broken valve springs. If the needle range fluctuates rapidly from 12 to 24 in-Hg (41 to 81 kPa), and fluctuation speed increases along with engine speed, the valve springs are weak, figure 4-10. A broken spring causes a rapid, irregular needle fluctuation each time the valve tries to close.

Ignition timing diagnosis

Suspect incorrect ignition timing when vacuum readings are steady but 2 to 3 inches (50 to 75 mm) above or below specifications.

Late ignition timing causes low gauge reading at idle, and early timing results in idle readings slightly above normal, figure 4-11.

Exhaust restriction diagnosis

With a vacuum gauge and tachometer connected, slowly open the throttle to bring engine speed up to 2000 rpm with the engine under no load. The needle reading should be approximately the same for idle and 2000 rpm. Quickly close the throttle and the needle should rise immediately to above normal, then return to normal. If the exhaust system is restricted, the vacuum reading will be lower at 2000 rpm than at idle, and the vacuum will not rebound when you quickly shut the throttle. A restricted exhaust system allows the vacuum to increase slowly when the throttle is first closed. Verify by holding engine speed at 2000 rpm while mon-

Figure 4-10 Weak valve springs produce rapid needle fluctuations that increase with engine speed.

itoring the gauge. If vacuum drops to near zero, then rises slowly to level off at a below normal reading, the exhaust is plugged, figure 4-12. Inspect for a frozen manifold heat control valve, a damaged or clogged muffler or catalytic converter, or a restricted exhaust pipe.

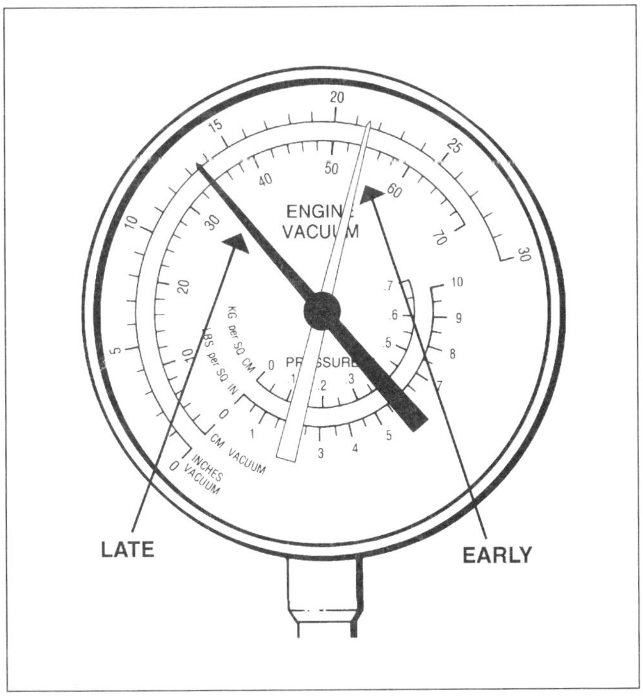

Figure 4-11 Late timing causes low readings with a steady needle, early timing causes high steady readings.

Figure 4-12 Exhaust restrictions cause the needle to drop, then level off below normal at high rpm.

Piston ring diagnosis

Before testing for leaking piston rings, make sure the engine has shown normal readings on all other tests. The crankcase oil must also be in good condition, diluted oil causes incorrect test results.

Snap open the throttle and accelerate the engine to 2000 rpm. When engine speed stabilizes, quickly close the throttle while watching the gauge. Good piston rings will cause the needle to jump 5 in-Hg (17 kPa) or more above normal, anything less indicates a compression loss.

Vacuum Control Devices

Vacuum is used to operate a wide range of systems and devices on late-model cars. In general, vacuum control devices can be divided into three groups:

1. Diaphragms (often called actuators or motors) use vacuum to operate a mechanical component, figure 4-13.
2. Switches, whether electric or thermostatic, change state to direct the flow of vacuum, figure 4-14.
3. Delay valves use a metered restriction to delay the flow of vacuum to a device, figure 4-15.

Vacuum diaphragm diagnosis

Any vacuum diaphragm should be able to hold a steady reading for at least one minute if it is in good condition. Test with a hand pump by applying 15 to 20 in-Hg (51 to 68 kPa) of vacuum to the diaphragm inlet,

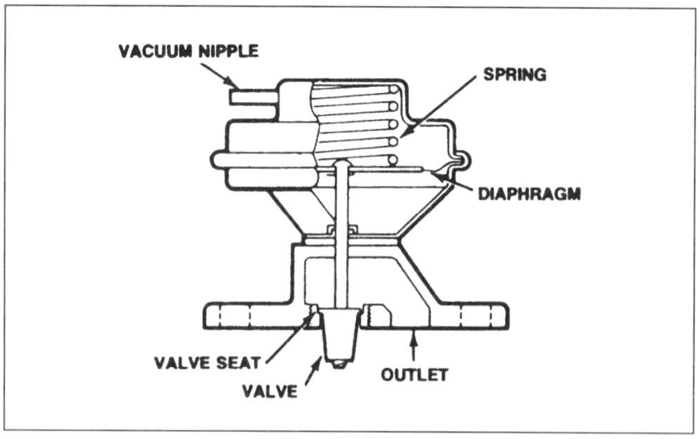

Figure 4-13 A diaphragm, like this EGR valve, uses vacuum to oppose spring pressure and perform a mechanical task.

figure 4-16. If the unit contains a bleed valve or clean air purge, it must be plugged during testing. Linkage attached to the unit must move smoothly without binding. Replace the unit if the diaphragm cannot hold vacuum.

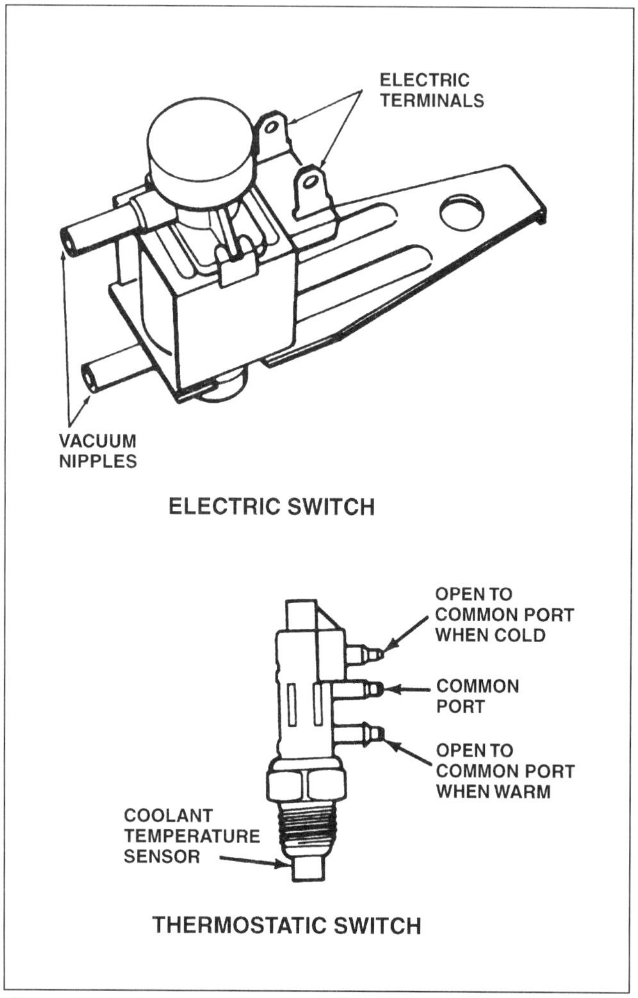

Figure 4-14 Electric and thermostatic switches are used to direct vacuum flow.

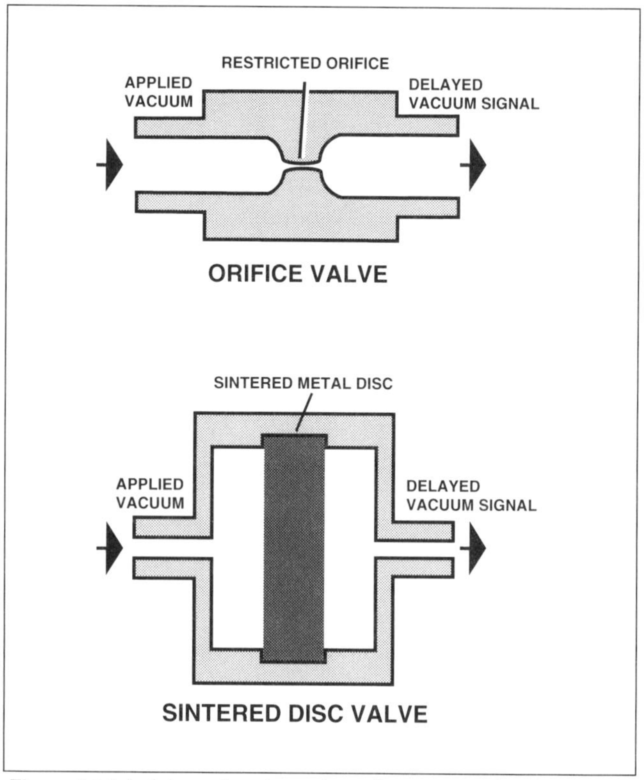

Figure 4-15 A delay valve slows down the vacuum signal.

Electrical vacuum switch diagnosis

Vacuum valves are often controlled by an electric solenoid, and are used to regulate vacuum flow to emission control devices. The solenoid moves a plunger to allow or prevent vacuum flow through the valve. Because a wide variety of designs are in use, you must follow the test procedure recommended by the manufacturer exactly. Some valves are normally closed and will hold vacuum when the solenoid is de-energized. Other valves may be normally open and cannot hold vacuum when the solenoid is de-energized.

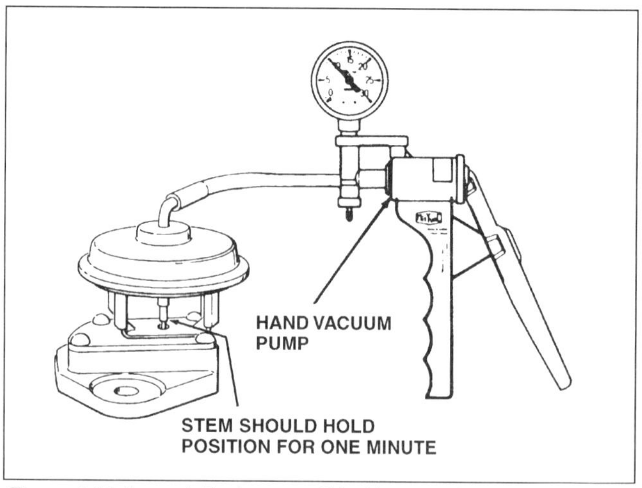

Figure 4-16 A good diaphragm will hold a specified vacuum for at least one minute.

Energizing the solenoid will cause it to change state, figure 4-17. If the unit was holding vacuum previously, it should release when power is applied. If the unit would not hold vacuum previously, it should when the solenoid is energized. A good solenoid will be able to trap and hold 15 to 20 in-Hg (51 to 68 kPa) of applied vacuum. If vacuum leaks down, the solenoid is defective.

Thermostatic vacuum switch diagnosis

Thermostatic vacuum switches also trap and hold applied vacuum as the switch changes states. However, state changes are induced by either coolant or air temperature. A variety of switch designs are used, so accurate specifications and test procedures are needed.

Typical test procedures are similar to those for electrical switches. When procedures require you test at high temperature, do not allow the engine to overheat. This is particularly important on cars with catalytic converters.

Delay valve diagnosis

Vacuum delay valves simply restrict, or slow down the vacuum signal. Test by connecting a hand pump to the inlet side and a second vacuum gauge to the outlet side of the valve, figure 4-18. Apply the specified amount of vacuum and note how many seconds it takes for the gauge readings on both sides of the valve to stabilize. If more or less time is required than specified, the valve is defective.

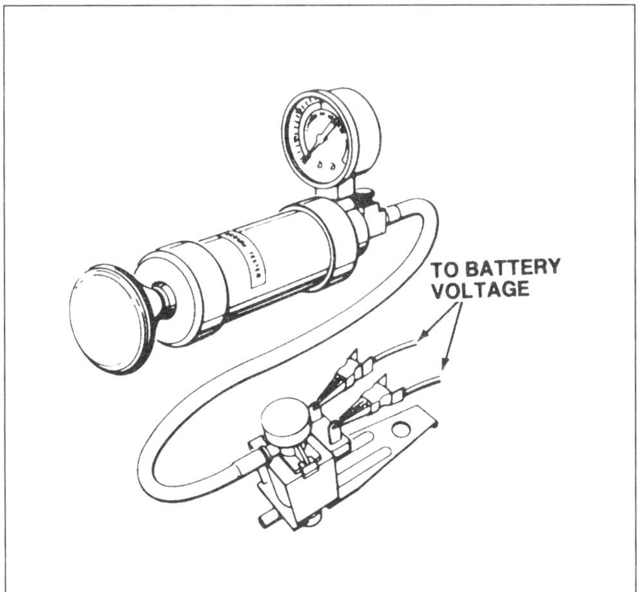

Figure 4-17 Electric vacuum valves switch state when connected to power.

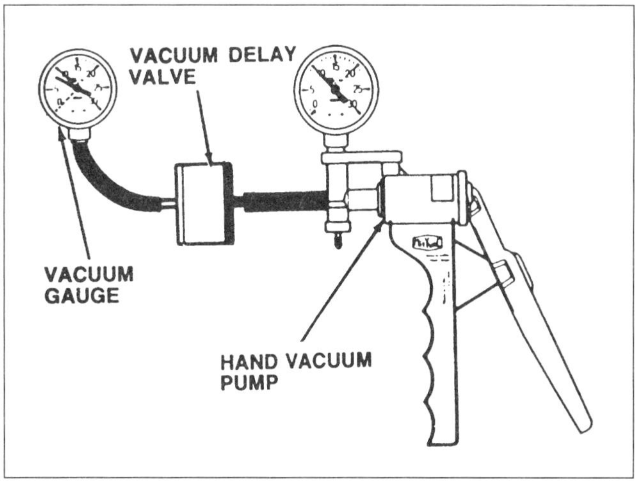

Figure 4-18 Vacuum delay valve test set up.

FUEL DELIVERY SYSTEM

To provide rapid engine startup, a fuel injection system retains a considerable amount of pressure when the fuel pump is not running. Follow procedures specified by the manufacturer to relieve this residual pressure before opening any fuel line for testing or service.

On an injection system, the malfunction of several components can result in a single symptom, or multiple symptoms can be caused by a single component. If you are to be efficient and effective in your troubleshooting, you must know and understand the operation of every input sensor and output actuator the system uses. Most of this information can be found in Chapter 2 of this book.

Fuel Delivery Problems

The first step is to check the fuel lines, both supply and return, and the fuel filter for possible restrictions. Next, verify that fuel pump delivery pressure is satisfactory. The final step is to check the operation of the fuel pressure regulator. You must have accurate specifications, and test according to the vehicle manufacturers' specific instructions. Following are general guidelines for troubleshooting the fuel delivery system.

No fuel system, regardless of type, will work properly if fuel pressure is incorrect. Injection systems precisely control fuel system pressure to ensure the correct amount of fuel is delivered to the engine. The composite vehicle system maintains regulated fuel system pressure between 34 to 38 psi (234 to 262 kPa), which is fairly typical for a multiport system.

Fuel pump diagnosis

Electric fuel pumps may be either the in-line or the in-tank type. In-tank units are often combined with the fuel gauge sending unit as an assembly, figure 4-19.

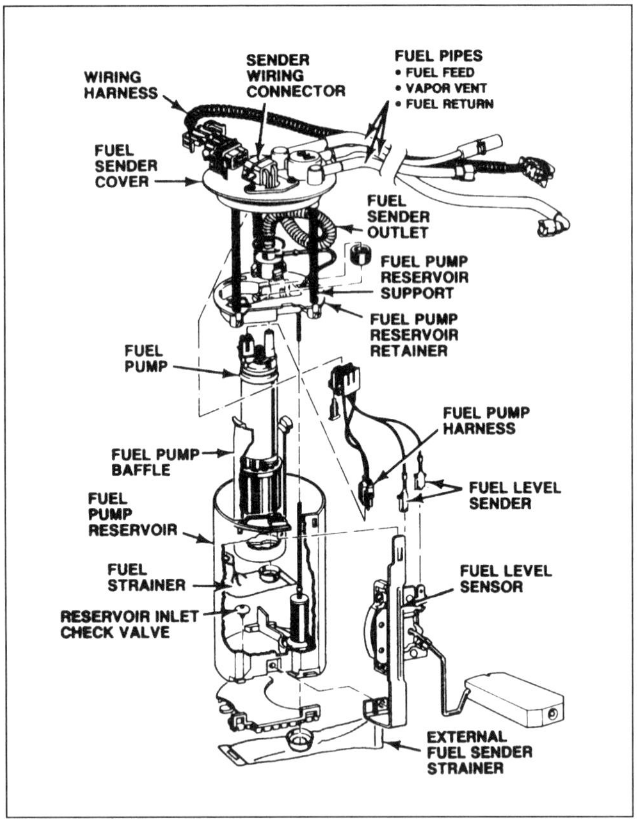

Figure 4-19 In-tank pumps are often combined in an assembly with the fuel gauge sending unit.

Most systems use a positive-displacement pump, which can be a vane, turbine, or roller design.

Since the pump is an electric motor, it needs a steady voltage signal and a good ground to operate. The power supply is generally through a fused circuit regulated by a PCM controlled relay. Fuel pumps are non-serviceable items, you replace rather than repair them.

Fuel pump relay diagnosis

Fuel pump relays can operate in one of two ways. The relay may be normally grounded and the PCM sends B+ to the relay to energize it, or the relay may receive constant power and the PCM grounds the circuit to energize the relay, figure 4-20. When the key is first turned on (with the engine off) the relay is usually energized for 1 to 2 seconds. If the PCM does not see a distributor or crank signal pulse, the relay will be de-energized and B+ will be removed from the fuel pump. Fuel pump relay information may show up in a data stream as a digital ON/OFF parameter.

Fuel pump relay problems

The following are examples of some of the driveability problems that can occur with fuel pump relay circuit problems.

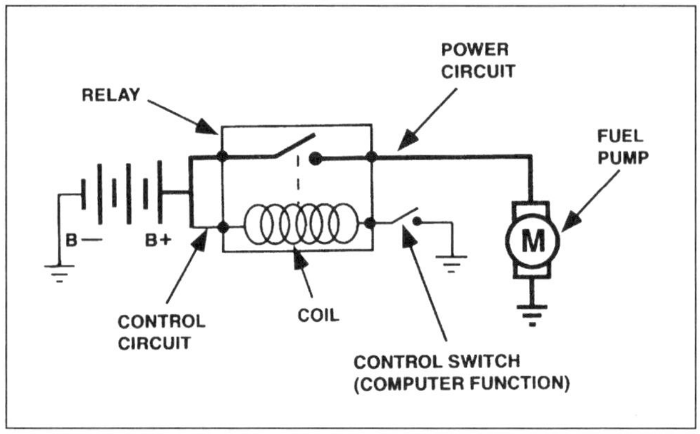

Figure 4-20 A fuel pump relay controls the power circuit to the fuel pump.

- A faulty PCM that does not energize a fuel pump relay after it has received a crank or reference signal will not allow the engine to start.
- An intermittent or faulty fuel pump relay can cause engine cut-out and possibly not allow engine starting.

Fuel system pressure diagnosis

Test fuel pressure with a gauge to determine if it is out of specifications, figure 4-21. You can also use the pressure gauge to locate restrictions that cause a pressure drop by measuring fuel pressure at various points in the system. The electronic control system compensates for small pressure changes, and monitoring the data steam can tell you how well the system is responding, figure 4-22. Driveability problems and emissions failures can be caused by both high and low fuel system pressure. Of the two problems, low fuel pressure is most common.

Low fuel pressure problems

With low pressure, the injectors do not deliver enough fuel to support efficient combustion, and the result is a lean running engine. Lean conditions cause the oxygen sensor signal to stay low, and the PCM responds by increasing injector pulse width. When pressure is

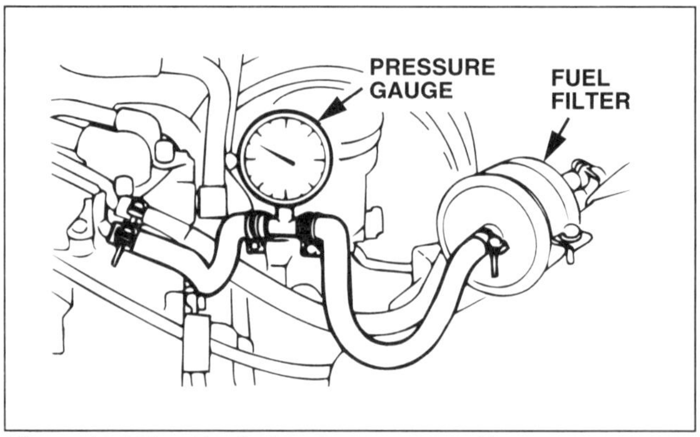

Figure 4-21 Checking fuel system pressure with a gauge.

too low, even the increased injector on time cannot keep up with the demands of the engine. Symptoms of low fuel pressure range from hesitation and lack of performance, to lean misfire and failure to start. These symptoms may be accompanied by oxygen sensor diagnostic trouble codes and, possibly, misfire codes on OBD II vehicles.

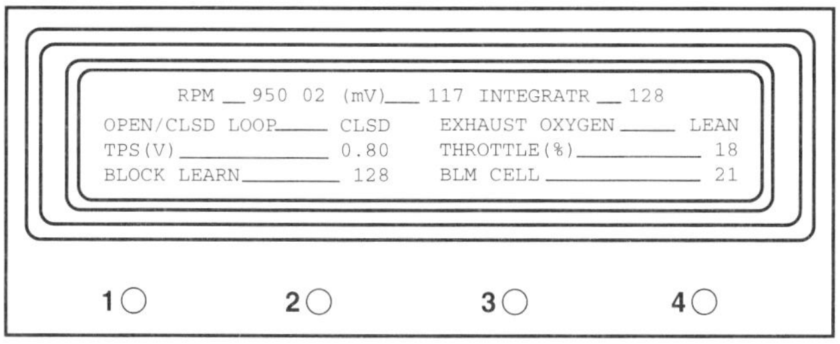

Figure 4-22 Fuel correction parameters shown on the serial data stream, such as Integrator (short term fuel trim) and Block Learn (long term fuel trim), can reveal some pressure problems.

Sources of low fuel pressure include:

- A clogged or restricted fuel filter or fuel tank pick-up
- A bent, kinked, or clogged fuel line
- A defective or malfunctioning fuel pump
- A defective or malfunctioning fuel pressure regulator
- A delivery system leak.

High fuel pressure problems

With high pressure, the injectors deliver too much fuel and the result is a rich mixture. In this case, the oxygen sensor signal should go high, causing the PCM to respond by reducing pulse width to lean the mixture. However, extreme high pressure can quench combustion flooding the exhaust with raw fuel to artificially raise the oxygen content. The O2S sees this as a lean condition and the PCM increases pulse width compounding the problem.

Symptoms of high fuel pressure can include surging, stalling, misfire, hard or no starting, and poor fuel mileage. Since high pressure results in a rich mixture, exhaust emissions will be high as well. Extended operation can lead to other problems, such as spark plug fouling, catalytic converter failure, and internal engine damage.

The cause of high fuel pressure will almost always be on the fuel return side, downstream of the injectors. Look for signs of kinking or other damage that can restrict flow through the return line to the tank. Also check the operation of the fuel pressure regulator. If the regulator is stuck closed, the system can not relieve excess pressure.

Fuel pressure regulator diagnosis

Fuel injection systems require a pressure regulator to maintain uniform fuel pressure at the injectors. Many use a hydro-mechanical bypass device with a spring-loaded diaphragm assembly to open a fuel return line to the tank when pressure exceeds system specifications, figure 4-23. Pressure regulators on most current systems install on the return (downstream) side of the injectors. This location minimizes pulsations caused by

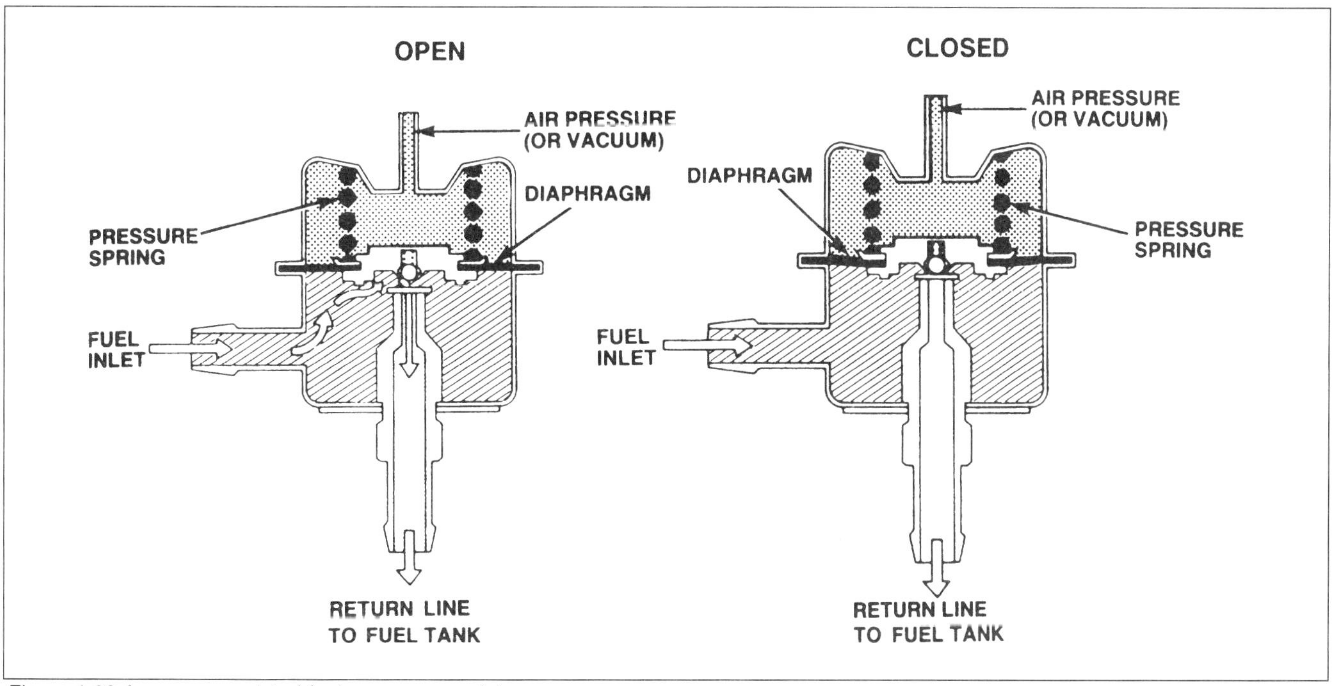

Figure 4-23 A vacuum assisted fuel pressure regulator.

pressure drop across the injectors as the nozzles open, ensures positive fuel pressure at the injectors at all times, and holds residual pressure in the lines when the engine is off.

Fuel pressure regulator problems

An out of specification pressure regulator can cause hard starting and numerous other problems. Fuel return to the tank is blocked if a regulator fails in the closed position, so system pressure will increase. If a regulator fails in the open position, too much fuel flows back to the tank and system pressure drops.

Some regulators are entirely mechanical and simply use fuel pressure to overcome spring pressure on the diaphragm and open the return line to the tank. Others are vacuum assisted and modulate fuel pressure through a combination of spring pressure and manifold vacuum acting on the diaphragm. Vacuum units must be able to hold vacuum applied with a hand pump without bleeding down.

Leaking fuel pressure regulator diaphragms can allow manifold vacuum to pull fuel past the diaphragm into the intake system, causing a rich condition. Symptoms include a higher than normal O_2 reading, fouled spark plugs, and an engine that will not start.

AIR INDUCTION SYSTEM

A running engine demands a constant supply of fresh intake air. On electronic injection systems, intake air must be precisely measured for the PCM to calculate the correct amount of fuel needed to meet the demands of the engine, figure 4-24. The intake air system can be divided into two sections, upstream and downstream. Upstream would be anything installed before the airflow sensing device, such as the air filter, thermostatic control valve, and ductwork. The downstream section includes the intake manifold, throttle body, ductwork, and any other components that connect the airflow meter to the engine. Each section presents unique concerns.

Upstream Intake Air

For the PCM to control fuel delivery for efficient combustion, intake air must be provided in sufficient quantity, and at the correct temperature. Problems in the upstream airflow can generally be traced to a restriction or a malfunctioning thermostatic control valve. Either condition can cause driveability concerns, but will seldom set a diagnostic trouble code.

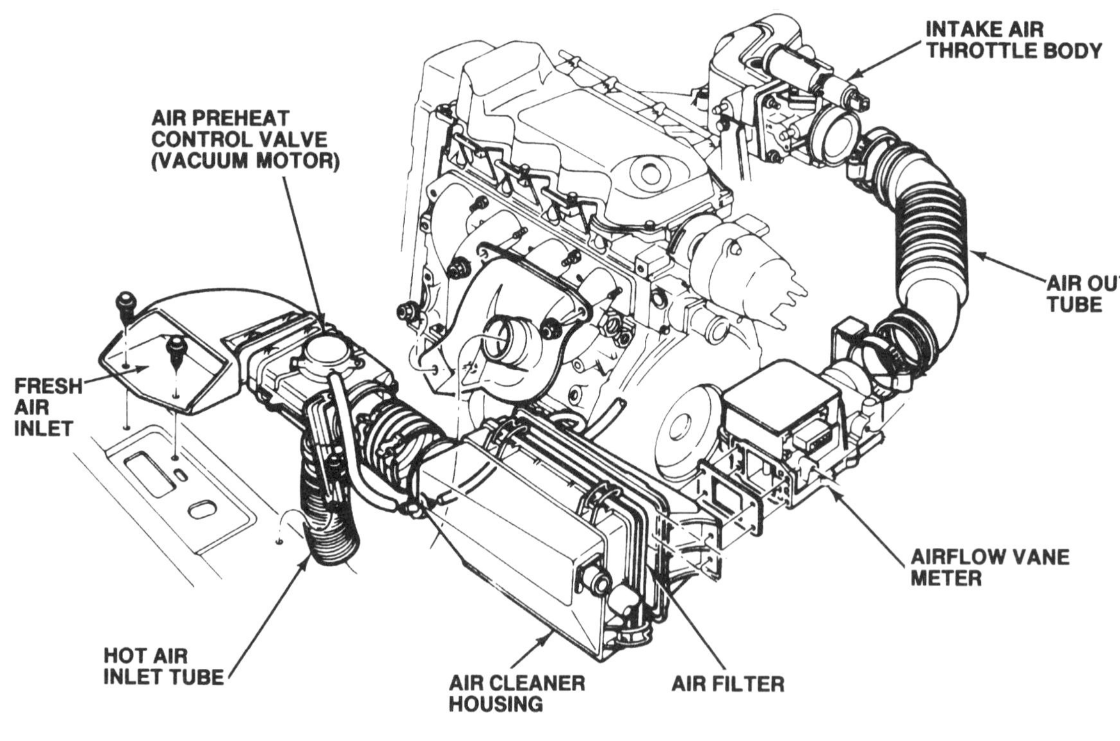

Figure 4-24 The entire intake air system must be in sound condition for good engine performance.

Air flow problems

A clogged air filter element is the most common source of an upstream air restriction. Since a restriction limits the amount of air allowed into the engine, it can result in a rich running condition for a carbureted engine or a lack of power for carbureted or fuel-injected engines. Inspect and replace the filter element as needed. Oily deposits on the element or sludge build-up in the filter housing indicate a problem in the PCV system. Inspect the housing and ductwork. Repair or replace parts that are broken, collapsed, or otherwise damaged.

Typical symptoms of restricted air flow include: hard starting, surging, misfire, lack of power, and poor fuel economy. Restrictions that create a rich running condition will cause abnormal activity on the serial data stream. Look for an oxygen sensor signal that stays high and switches slowly. Injector pulse width and fuel correction parameters should be low as the PCM attempts to lean out the mixture.

Air temperature problems

Engine operation is upset when the intake air thermostatic control does not work correctly. A cold engine requires warm intake air for efficient combustion. When intake air temperature is too low, engine performance is erratic. Typically, the engine tends to stall easily and will not accelerate properly.

Once the engine is up to temperature, the air control valve must close off the preheated air to maintain operating efficiency. High intake air temperature can cause a power loss, and possibly detonation. Typical complaints are an engine that seems to be running out of fuel, and rough, or erratic automatic transmission shifting.

On the composite vehicle, and many other systems, the PCM calculates intake air temperature (IAT) using the variable voltage signal from a thermistor-type sensor. The data stream may include parameters for temperature, voltage, or both. Only the IAT sensor signal voltage is available when scanning the composite vehicle. This is a typical 5-volt circuit, so 2.5 volts is the midpoint, readings should be near this point under normal operating conditions. When intake air temperature is 86°F (30°C), IAT signal voltage is 2.6 volts on the composite vehicle, figure 4-25. Signal voltage is high when temperature is low, and goes down as temperature comes up.

Intake air control valves

A blend door is used to regulate cold and preheated air. Vacuum diaphragms, temperature sensors, and thermostatic bulbs are used to move and position the blend door. The primary purpose of the air control valve is to reduce hydrocarbon (HC) and carbon monoxide (CO) emissions from cold engines. Malfunctions can cause emission test failure.

Vacuum diaphragm problems

Generally, applying vacuum will open the preheated air passage. So, cold driveability suffers because there is no warm air available when the unit fails.

Temperature F	(C)	Sensor Voltage
248	(120)	0.25
212	(100)	0.46
176	(80)	0.84
150	(66)	1.34
140	(60)	1.55
104	(40)	2.27
86	(30)	2.60
68	(20)	2.93
32	(0)	3.59
-4	(-20)	4.24
-40	(-40)	4.90

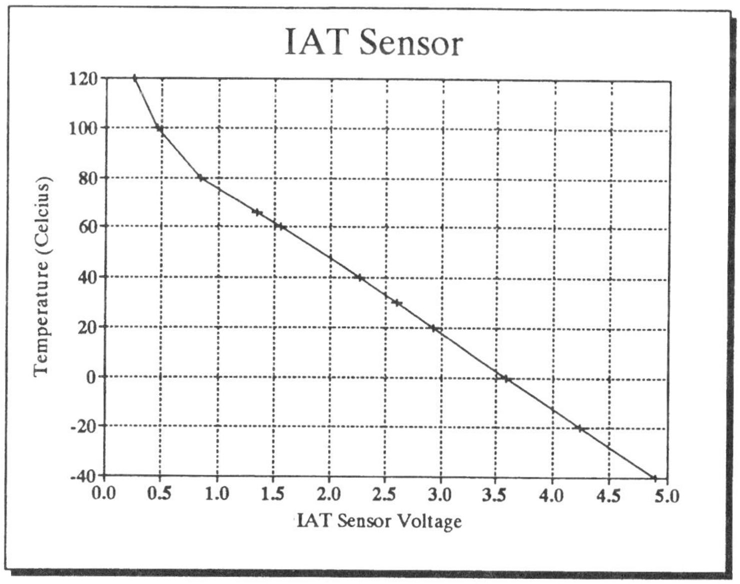

Figure 4-25 Composite vehicle IAT sensor performance specifications.

Temperature sensor problems

Sensor specifications for when the blend door starts to open, and for when it is fully open are generally provided by the manufacturer. Sensors often incorporate a switch used to regulate vacuum flow to a diaphragm, figure 4-26. Use a thermometer to check temperature as you start and run the engine. As with a diaphragm, preheated air is unavailable when the unit fails. Poor cold driveability and high emissions are typical symptoms.

Thermostatic bulb problems

Checking the function of a thermostatic bulb, figure 4-27, requires a bench test. The hot air inlet should close off as the unit soaks in a hot water bath. Typically, it will take water temperature of about 130° to 150°F (53° to 65°C) to fully close the preheat damper.

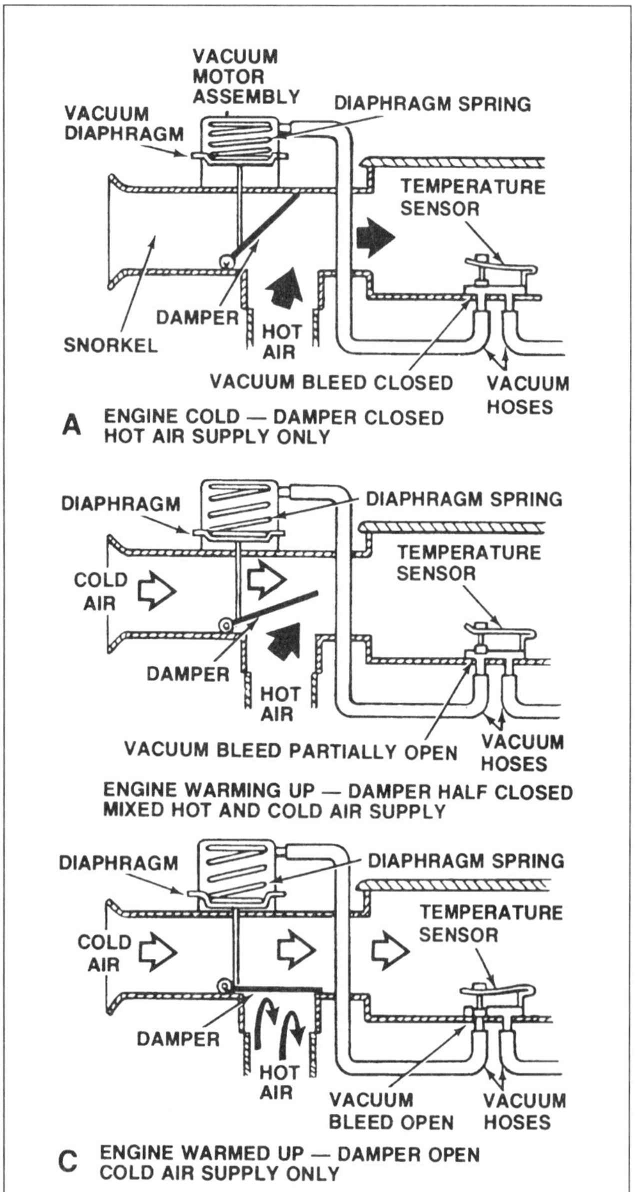

Figure 4-26 A temperature sensitive switch used to activate a blend door diaphragm.

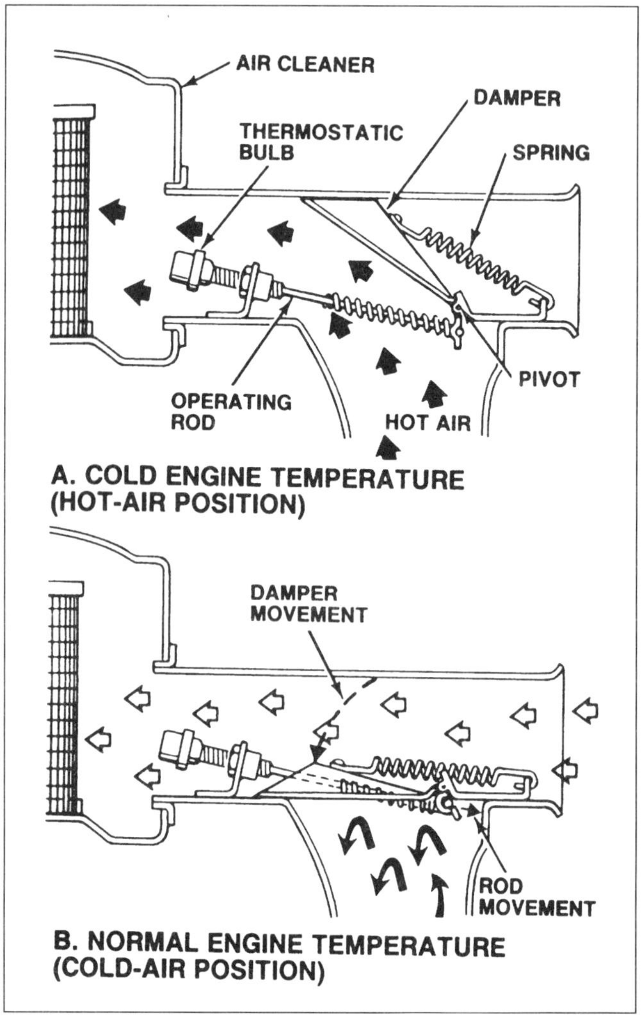

Figure 4-27 Thermostatic bulb assembly requires a bench test.

Manifold heat control valves

On some vehicles, intake air is preheated by circulating warm air off the exhaust or coolant through intake manifold passages. Spring pressure, figure 4-28, or vacuum, figure 4-29, can be used to regulate preheat airflow. A thermostat is used to control flow on systems that operate with coolant. With any design, a valve stuck open can cause increased fuel consumption, poor performance during warmup, and increased exhaust emissions. A valve stuck closed can cause poor acceleration, a lack of power, and poor high-speed performance.

Downstream Intake Air

On a mass airflow system, any air leaking into the engine downstream bypasses the airflow meter and cannot be detected by the PCM. There is no fuel correc-

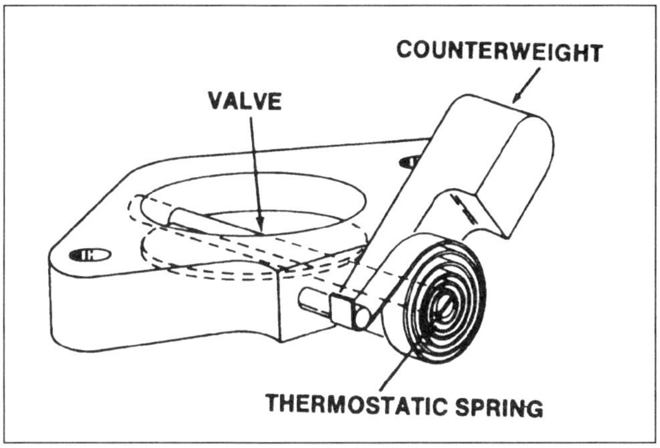

Figure 4-28 Mechanical spring-tension thermostatic heat control valve.

tion to compensate for the additional air since the PCM does not see it. As a result, the air-fuel mixture goes lean. This sort of leakage is known as unmetered air.

Injector pulse width may increase after the fact in response to a low oxygen sensor signal. However, mixture is usually unstable and driveability is effected. Severe leaks can exceed the correction capability of system, prevent O2S switching, and cause the system to remain in open loop.

On a speed density system, any air leaks affecting vacuum to the map sensor cause lower than expected vacuum readings on the sensor. This causes the map sensor to send an incorrect signal to the PCM, indicating a heavier load than is actually on the engine. This false signal causes the fuel system to run richer than is actually needed.

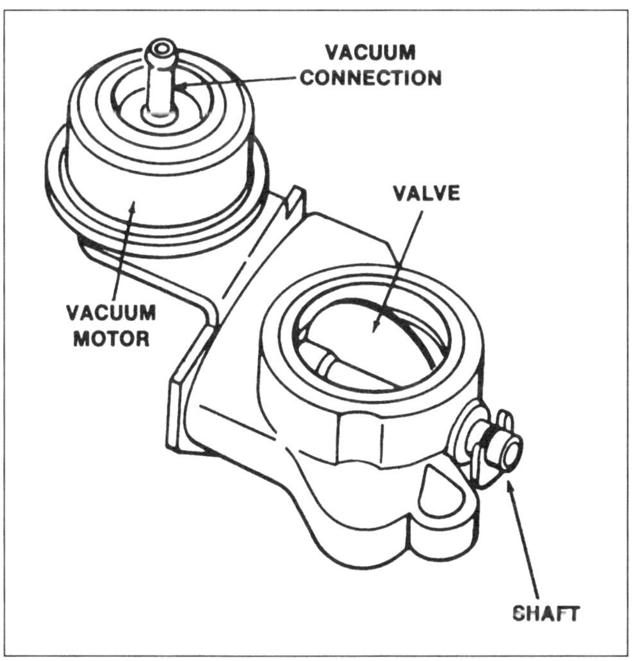

Figure 4-29 Vacuum actuated thermostatic heat control valve.

Symptoms of an unmetered air leak are usually a high or unstable idle speed, and possibly stalling. Inspect intake air ductwork for leaks, and also check the condition and routing of all vacuum, breather, and bypass hoses. Most leaks can be detected through manifold vacuum testing as previously described in this chapter.

Idle air control (IAC) valve diagnosis

The idle air control (IAC) valve on most systems is a variable position stepper motor. The motor regulates the flow of air that bypasses the throttle plate with a pintle valve to maintain a constant idle speed, figure 4-30. Malfunctions in the idle speed control system may or may not set a diagnostic trouble code. Check for loose connections and damaged wiring to the idle control motor and all of the input sensor circuits. Input sensors that affect idle speed can include some or all of the following:

- Throttle position
- Coolant temperature
- Airflow
- Manifold absolute pressure
- Engine speed
- Vehicle speed
- Exhaust gas oxygen
- Exhaust gas recirculation
- Canister purge
- Air conditioner on/off
- Cooling fan on/off

Some IAC motors are basically pulse-width modulated solenoids, so the signal on a running engine can be checked with an oscilloscope or voltmeter. Check continuity of the solenoid windings with an ohmmeter. Many systems, like the composite vehicle, transmit IAC position information as a data stream parameter. Data

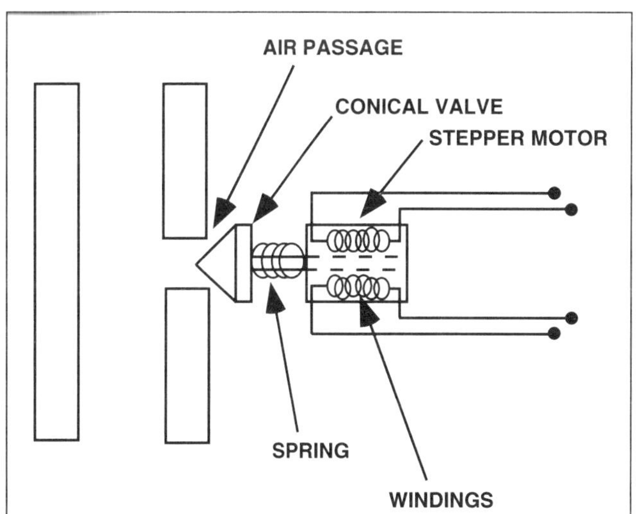

Figure 4-30 A stepper motor maintains a constant idle with incremental adjustments to bypass air flow.

may be presented as a percentage, or a numerical count. Zero percent or a low count indicates the bypass is closed, and a high count or a 100 percent reading indicates the bypass passage is fully open.

Symptoms of an IAC failure include a high, low, or erratic idle speed, and possibly a tip-in sag, hesitation, and deceleration stalls.

Throttle body diagnosis

Throttle body problems include air leakage from worn throttle shaft bores or poor base gasket sealing, and an unstable air supply due to sludge deposit build-up on the throttle plates, housing bore, and IAC valve. Both conditions can cause driveability complaints and emissions test failures.

Air leaking at the throttle body base or throttle shafts will lower manifold vacuum. Expect vacuum gauge readings to be below normal with a fluctuating needle. Idle speed and quality may be effected as well.

Detecting sludge deposits, or throttle body coking, may require a visual inspection. However, data stream parameters can point you in the right direction. IAC values will tend to be high as the system attempts to let in more air to compensate for the restriction caused by the deposits. Symptoms of throttle body coking are most noticeable at idle, off-idle acceleration and deceleration. Complaints include erratic or unstable idle speed, tip-in sag, hesitation, stalls, and poor performance.

FUEL INJECTION-SYSTEM

On a modern engine, the fuel injection must be considered as a subsystem of the complete engine control system. Malfunctions on any of the other subsystems may mimic a fuel-injection failure. The injector nozzles are the last step in a long chain of electronic, hydraulic, and pressure commands. Problems anywhere along the way can cause an apparent injector malfunction, so you must eliminate these possibilities first. An injection system must adjust the air-fuel ratio to meet the demands of different operating modes. The injector nozzles ultimately correct the air-fuel ratio by increasing or decreasing the amount of fuel delivered to the combustion chamber. Electronic controls are discussed in Chapter 2 of this book. This chapter address the mechanical operation of the injectors. To review basic engine air-fuel requirements:

- On startup — Low airflow and manifold vacuum, combined with a cold engine and poor fuel vaporization, require a rich air-fuel ratio
- At idle — Low airflow and high manifold vacuum, combined with poor vaporization, still require a slightly rich air-fuel ratio
- At low speeds — Air-fuel ratio becomes progressively leaner as engine speed, airflow, and vacuum increase
- At steady cruising — Air-fuel ratios become the leanest for best economy with light load and high, constant vacuum and airflow
- On acceleration — Air-fuel ratio is rich and combined with low vacuum and high airflow
- Under heavy load — A rich air-fuel ratio is combined with low vacuum and high airflow at wide-open throttle.

Two types of injection nozzle, mechanical and electronic, are used to deliver fuel to the combustion chambers. On late-model systems, electronic injectors are more common. We briefly discuss both types here.

Mechanical Fuel Injectors

Mechanical fuel injectors operate on the opposing forces of fuel pressure and a spring-loaded valve in the injector tip, figure 4-31. The most common example used on production vehicles is the Bosch K-Jetronic continuous injection, or CIS system.

Typically, when inlet fuel pressure reaches about 45 psi (310 kPa), it overcomes spring pressure and forces the injector open. All of the injectors deliver fuel continuously whenever pressure is above the established threshold. These systems operate on a variable pressure basis to regulate the amount of fuel delivered. Injector nozzle opening increases along with pressure to deliver more fuel on demand. Injectors are located in the port near the intake valve, and fuel collects behind the valve until the valve opens. As the valve opens, turbulence of the incoming air atomizes the fuel as the mixture is drawn into the combustion chamber.

With CIS, the main fuel distributor and regulator controls fuel pressure in response to airflow and preset regulator pressure. Injector spring force and fuel pres-

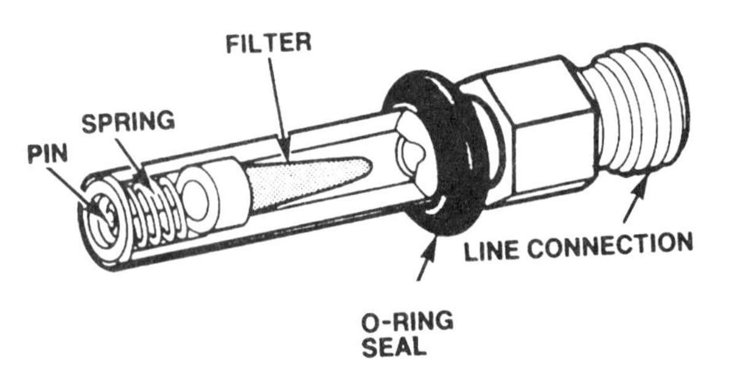

Figure 4-31 Mechanical injectors open when fuel pressure exceeds spring pressure.

sure cause the injector valve to vibrate. This constantly varies the amount of fuel in relation to engine requirements and helps to atomize the fuel as it is injected. When the engine is shut off, fuel pressure drops below injector tip spring pressure, and the injectors close. Residual fuel pressure is retained in the lines to ensure a rapid restart.

Hydraulically operated injectors can be tested for opening pressure, leakage, and delivery volume. They require the use of a special pressure tester, figure 4-32, and volume tester, figure 4-33.

Mechanical injector diagnosis

Hydraulically operated injectors are tested for opening pressure, leakage, and delivery volume. They must be removed from the vehicle for testing, and two special tools are required.

A pressure tester is used to check opening pressure, spray pattern, and leakage. Incorrect opening pressure causes the injectors to open early or late to result in either a rich or lean condition. Leakage generally causes a rich mixture accompanied by a pressure drop. The injector valve must close and seal tightly at specified pressure. Spray pattern must be conical and even, figure 4-34. A bad pattern disrupts air-fuel blending and causes incomplete combustion, which leads to increased emissions.

A volume tester measures how much fuel each injector delivers in a specific amount of time. Fuel volume problems can cause rough running, poor performance, and possibly misfire, as they disrupt cylinder balance. Emissions are also effected.

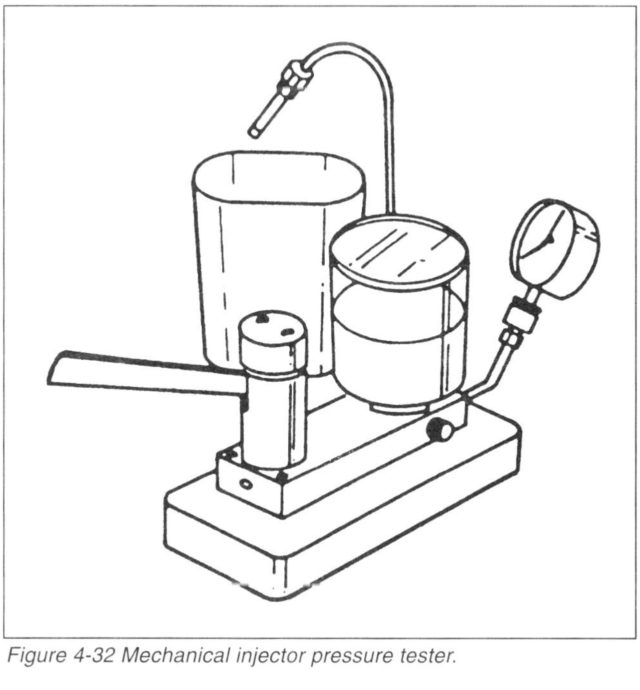

Figure 4-32 Mechanical injector pressure tester.

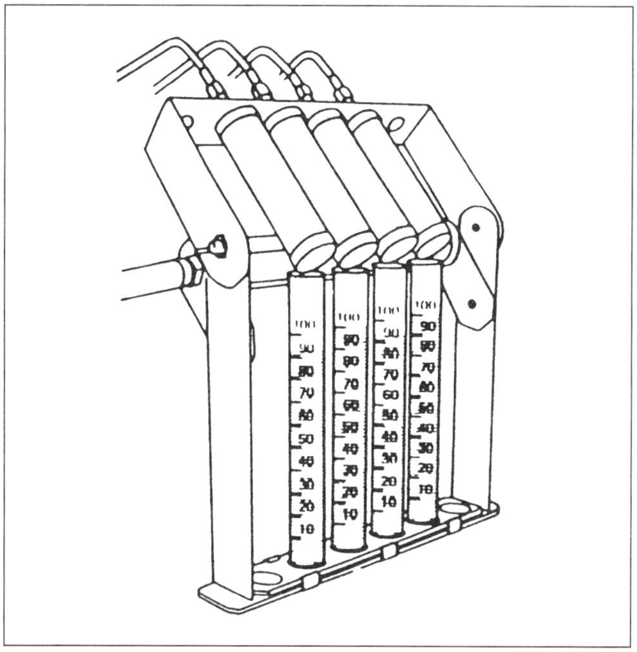

Figure 4-33 Mechanical injector volume tester.

Electronic Injectors

Electronic injection nozzles are pulse-width modulated solenoids that use either a needle or ball valve to open and close the fuel port, figure 4-35. Spring pressure holds the valve closed against its seat, and the solenoid armature opens the valve according to a pulsed signal voltage from the PCM. Generally, constant voltage is applied to the injector and the PCM regulates the ground side of the circuit.

Test electronic injectors as you would any other solenoid-operated device. Specifications are generally available for checking solenoid winding resistance with

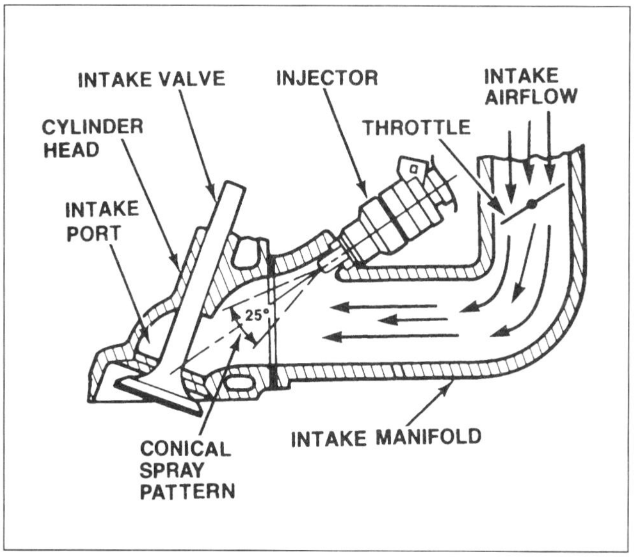

Figure 4-34 An even, conical injector spray pattern is essential for good fuel atomization.

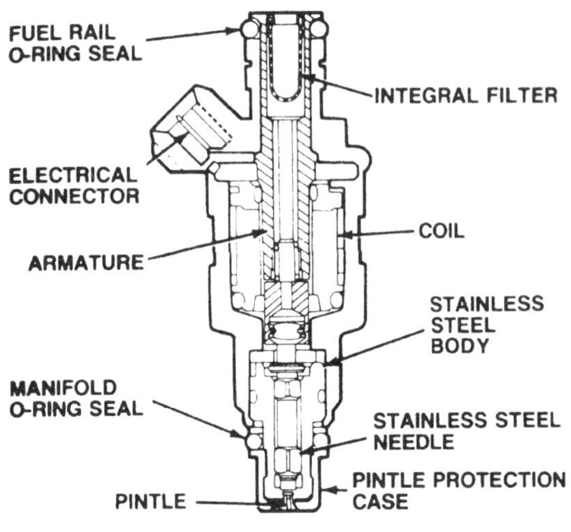

Figure 4-35 Electronic injectors are pulse width modulated solenoids.

an ohmmeter. Use a dwell meter to check duty cycle. Use a lab oscilloscope to monitor injector signal voltage activity, figure 4-36.

Electronic injector valves always open the same distance, the amount of time they remain on determines the richness of the air-fuel mixture. Some systems transmit injector parameters on the data stream, others, such as the composite vehicle, do not. Like their mechanical counterparts, electronic injectors must be able to hold pressure without leakage, and deliver sufficient fuel in a good pattern to support efficient combustion.

Weak or defective injectors often show up during a cylinder balance test. When spark and compression are good, an injector problem is indicated by uneven or no rpm drop during the test. Perform a cylinder balance test only when it is recommended by the manufacturer, the procedure can cause problems for some catalytic converter-equipped vehicles.

All injectors have a pressure drop specified by the manufacturer for testing on a flow balance tester. Flow balance testing is performed with the injectors installed in the engine, the fuel system pressurized, and the engine off. The equipment energizes each injector for a precise time period as you monitor fuel rail pressure drop, figure 4-37.

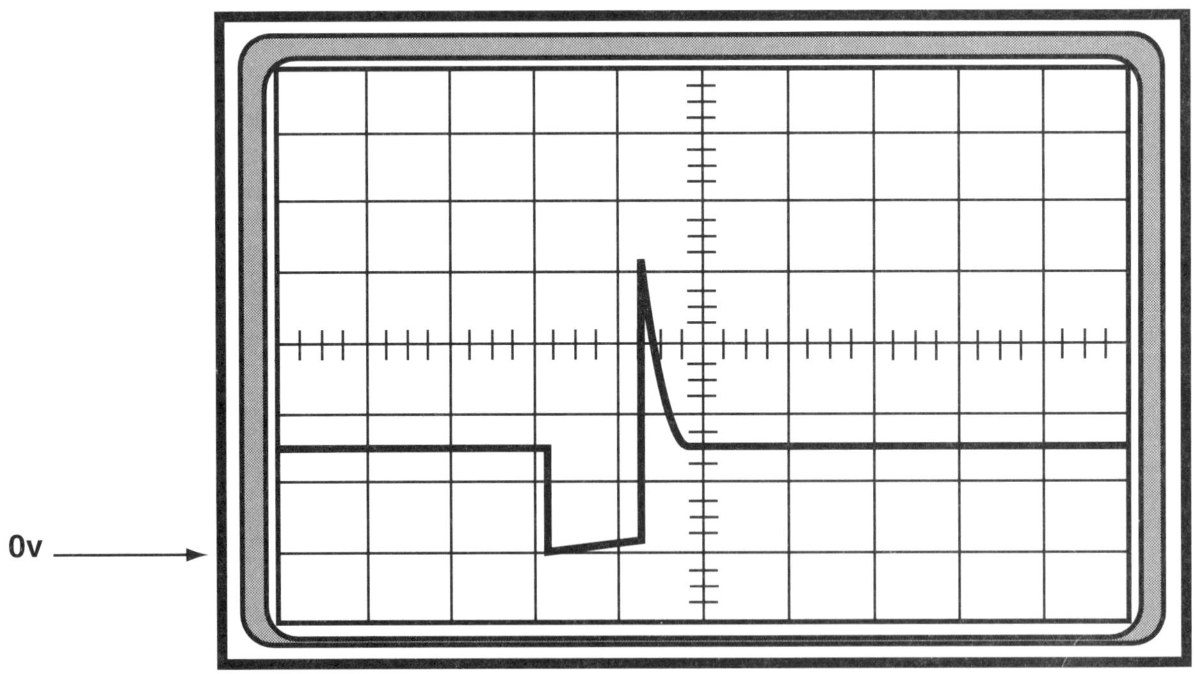

Figure 4-36 Typical oscilloscope trace of a good fuel injector.

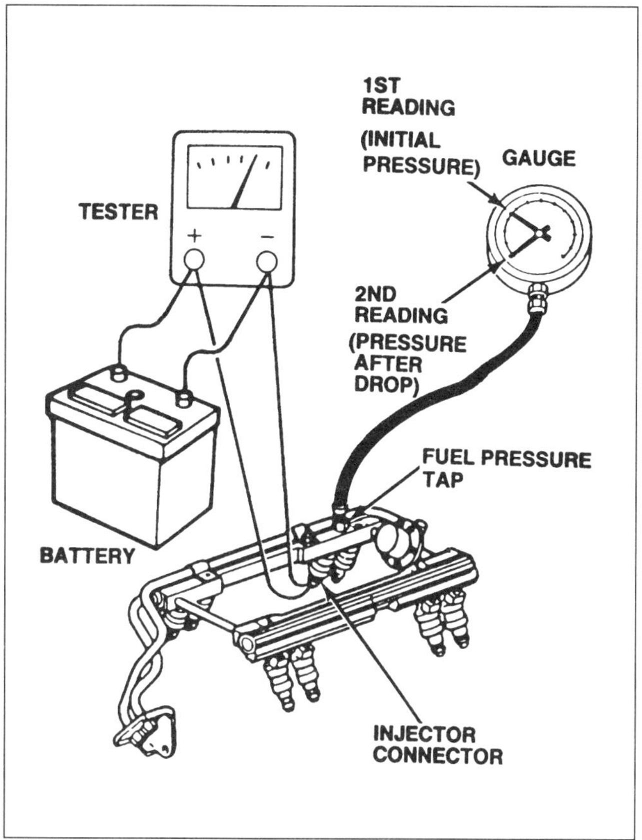

Figure 4-37 A flow balance tests measures pressure drop as the injectors are energized one at a time.

Electronic injector diagnosis

Injectors generally seal to the intake manifold with an O-ring, and air leaks here are common. One leaking O-ring can produce a rough idle and possibly a lean misfire from the offending cylinder. Multiple seal failures can cause an overall lean condition with erratic running, poor performance, and all of the other symptoms that go along with lean running.

A common injector problem is poor performance due to deposit accumulations. Deposits restrict the amount of fuel the injector can deliver leading to an incorrect air-fuel ratio, and also disrupt the spray pattern leading to poor atomization and incomplete combustion. Since either situation effects combustion efficiency, emissions test failure can result. In addition to high emissions, deposit contamination can cause hard starting, poor idle, surging, power loss, and poor fuel mileage.

Other injector mechanical failures include internal breakage and fatigue due to wear. This generally results in a loss of spring pressure, making the unit unable to hold back fuel system pressure. The resulting leakage delivers too much fuel to the cylinder, leading to an increase in emissions. Symptoms are similar to those of high fuel pressure and include hard stating, erratic idle, surging, stalling, misfire, and poor fuel mileage.

1. On a fuel-injected engine, the fuel pressure regulator vacuum hose is unplugged.

   Technician A says that this will cause the fuel pressure to increase.

   Technician B says that this will cause the fuel pressure to decrease.

   Who is right?

   a. A only

   b. B only

   c. Both A and B

   d. Neither A nor B

   Explanation: The vacuum hose to the control pressure regulator receives intake manifold vacuum to determine engine load. Having the vacuum hose unplugged would be simulating a full load condition requiring more fuel. This will cause the pressure to increase, so A is correct.

2. The PCM has a faulty injector driver circuit.

   Technician A says to replace the PCM.

   Technician B says to check the injector resistance.

   Who is right?

   a. A only

   b. B only

   c. Both A and B

   d. Neither A nor B

Explanation: Any time you replace a computer due to a faulty driver, you should check the resistance of the component that the driver was energizing. If the PCM driver failed due to a shorted injector, replacing the PCM will not solve the root of the problem. If the injector is not replaced, the PCM will most likely fail again, so both Technicians are correct.

3. The injector is displayed on a scope below, which of the following is true?

   a. This is a normal injector pattern.

   b. The injector driver has an open circuit.

   c. The injector driver is shorted.

   d. The PCM has a high resistance at its power ground.

   Explanation: The injector driver transistor in the PCM depends on the power ground circuit to turn the injectors on by pulling the signal low. With a high resistance at the power ground circuit, the injector driver was not capable of pulling the signal even close to ground. When a standard non current limiting injector is turned on, the injector will read a small amount of voltage during the time it is energized, but it should be less than one volt, so d is the best answer.

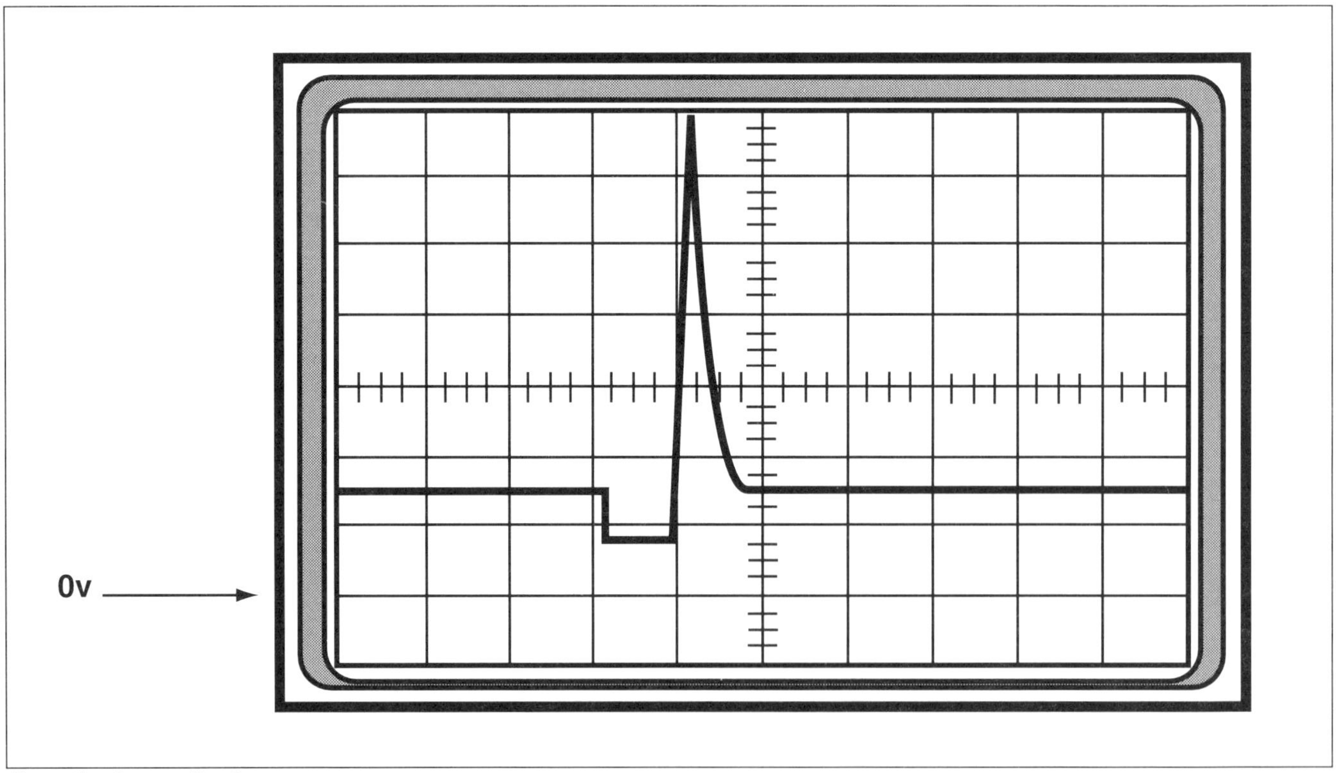

Illustration for question 3.

4. When the engine is off, the hot air passage remains open and the cold air passage remains closed in a vacuum-operated air cleaner. This could be caused by:
   a. High ambient air temperature
   b. A broken vacuum motor spring
   c. Low ambient air temperature
   d. The heat control valve stuck closed

   Explanation: Ambient temperature would not have an effect on a vacuum operated air cleaner. But the vacuum motor spring does keep the hot air passage closed to hot air. If the spring breaks, then it will not be able to force the hot air passage closed, so b is the correct answer.

5. Generally, a mechanical fuel pump should deliver a fuel volume of:
   a. One pint per 45 seconds in one minute at 500 rpm
   b. Two quarts per minute at 500 rpm
   c. One gallon per minute at 500 rpm
   d. One quart per five minutes at 500 rpm

   Explanation: Any mechanical fuel pump delivering at least a pint in 45 seconds is considered to be producing a normal volume, so a is the best answer.

6. Pressure and volume tests of a mechanical fuel pump are both below specs.

   Technician A says that an air leak in the fuel line between the tank and the pump could be the cause.

   Technician B says a plugged fuel tank pick-up could be the cause.

   Who is right?
   a. A only
   b. B only
   c. Both A and B
   d. Neither A nor B

   Explanation: Any restriction or displacing of fuel will reduce both pressure and the volume. Therefore, c is the best answer.

7. If the rpm gain measured during artificial enrichment tests is less than specified:
   a. The air-fuel ratio is too lean
   b. The engine is overheated
   c. The air-fuel ratio is too rich
   d. The air injection system is not working correctly

   Explanation: For the RPM gain to increase during the artificial enrichment test the fuel system must be leaned out, so c is the best answer. In other words, because adding more propane does not increase the RPM reading, the engine is receiving more fuel than it should in order to meet specs and must be leaned out.

8. In an electronic fuel-injection system, fuel pressure that is below specifications can be caused by:
   a. A defective fuel regulator
   b. Restriction in fuel lines or filter
   c. A defective fuel pump
   d. All of the above

   Explanation: All of the above items can cause low fuel pressure. Any restriction on the inlet side will reduce the fuel pressure.

9. The exhaust manifold heat riser is stuck in the open position.

   Technician A says this can cause poor gas mileage.

   Technician B says this can cause the intake manifold vacuum to be lower than normal.

   Who is right?
   a. A only
   b. B only
   c. Both A and B
   d. Neither A nor B

   Explanation: When the heat riser is stuck open during warm-up the engine will run poorly, take longer for warm up, and fuel will not evaporate well. As a result there will be wasted fuel, so Technician A is correct.

10. Technician A says low-pressure fuel-injection systems operate at 15 to 25 psi.

    Technician B says high-pressure fuel injection systems operate at 35 psi or more.

    Who is right?
    a. A only
    b. B only
    c. Both A and B
    d. Neither A nor B

    Explanation: Most low pressure fuel injection systems operate between 15 to 25 psi. Most high pressure systems usually operate at 35 psi or more. Both a and b are true.

Chapter Five

EMISSIONS CONTROLS SYSTEMS DIAGNOSIS

All new vehicles must meet government standards for both active and passive emissions before being certified for sale. Active emissions are those emitted from the tailpipe and tested by exhaust gas analysis. Passive emissions are toxic fumes or vapors, generally hydrocarbons, that rise off the vehicle while it is parked and not running. Although there are many passive emission sources, vapor from fuel in the tank and lines is the only source currently regulated by the vehicle control system. Largely ignored by the service industry in the past, both I/M 240 enhanced testing and OBD-II regulations require the monitoring of passive emissions systems.

Many components of the fuel, ignition, and engine management systems play a role in controlling vehicle emissions. Therefore, they must be considered part of the emission control system. The appropriate chapters of this book discuss many of these items in detail, so they will only be briefly covered here. The focus of this chapter will be on those subsystems and components whose primary function is to control emissions. Emission control system parts include:

- Positive crankcase ventilation
- Exhaust gas recirculation
- Ignition timing control
- Idle air control
- Catalytic converter
- Secondary air injection
- Intake air temperature control
- Early fuel evaporation control
- Evaporative emission control.

POSITIVE CRANKCASE VENTILATION (PCV)

Positive crankcase ventilation (PCV) systems recirculate the vapors and blowby gases produced in the crankcase back into the induction system of the engine. PCV system designs vary, but essentially; crankcase gases are drawn off at the valve cover and introduced into the intake charge through a PCV valve, figure 5-1, calibrated orifice, or separator.

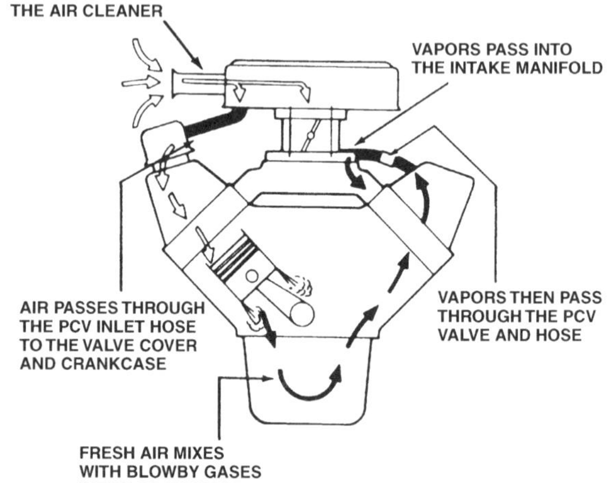

Figure 5-1 PCV system operation under normal operating conditions.

Crankcase vapor has a high hydrocarbon content, so when the crankcase vents, the air/fuel mixture goes rich. Because mixture is effected, PVC system problems can set diagnostic trouble codes and cause driveability problems.

PCV Driveability Problems and Emission Failures

For the PCV system to work properly, the flow control valve must function and allow air to flow freely through the system, figure 5-2. Before transmitting fuel control output signals, the PCM consults programmed lookup tables to compensate for the effect of PVC flow on the mixture. On some engines, as much as 30 percent of idle air is delivered through the PCV system. Flow problems, or excessive blowby vapor in the PCV system result in driveability problems and emission failures.

A PCV system that is not properly vented, or scavenged, causes oil dilution and sludge formation. With sludge, there are no immediate driveability symptoms. The long-term effect is accelerated engine wear and inefficient air cleaner operation, which leads to other problems and failures.

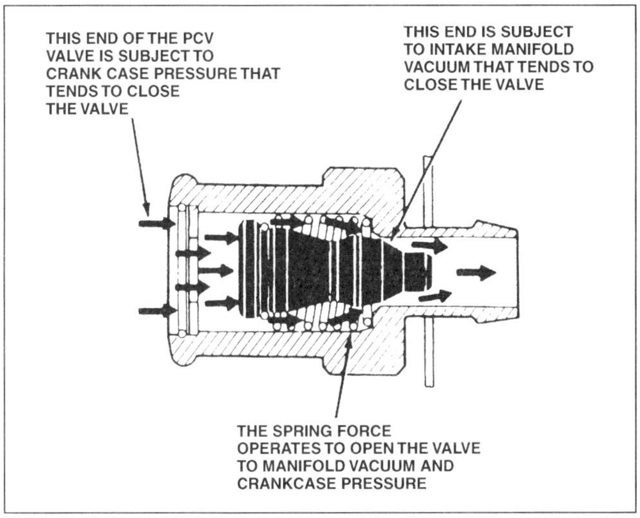

Figure 5-2 A combination of spring tension, crankcase pressure, and manifold vacuum regulate flow through a PCV valve.

When a PCV system becomes restricted or clogged, the cause is usually an engine problem or the lack of proper maintenance. For example, scored cylinder walls or badly worn rings and pistons will allow too much blowby. Start-and-stop driving requires more frequent maintenance and causes PCV problems more quickly than highway driving, as will any condition allowing raw fuel to reach the crankcase. Using the wrong grade of oil, or not changing the crankcase oil at periodic intervals will also cause the ventilation system to clog.

Restricted PCV flow

A partly or completely clogged PCV valve, or one of the incorrect capacity, may cause poor engine performance. A clogged PCV system may cause:

- Oil leaks due to increased crankcase pressures
- Increased oil consumption by the engine
- Diluted or dirty oil caused by accumulated sludge, water, and acids in the crankcase
- Escaping blowby vapors from the dipstick tube, oil fill tube, valve cover gasket areas, and any other opening to the crankcase
- Uneven or rich fuel mixture at idle and low speed. The PCM may set a rich fuel mixture DTC.
- Stalling or a rough idle

Excessive PCV flow

A PCV system with a vacuum leak or too great a flow rate may cause:

- Lean fuel mixture at idle and low speeds. The PCM may set a lean mixture DTC.
- Rough idle due to the lean mixture.

PCV valve problems

A replacement PCV valve that does not meet original equipment standards can cause problems. Driveability may be effected if the flow rate of the valve varies too much from that of the PCM look-up tables. Improper flow in the PCV system may upset the programmed values expected by the PCM for manifold pressure in relation to throttle angle, rpm, and barometric pressure. On some systems this can set a DTC for the MAP sensor. When the PCM relies heavily on measurement of manifold pressure for its fuel control programs, a variation of 10 to 15 percent in the MAP sensor signal may be enough to cause driveability problems, and possibly set a code.

PCV system information is not available on the serial data stream. However, related fuel correction parameters are, figure 5-3. Manipulating PCV flow while monitoring correction parameters is a good way to check PCM function and response.

EXHAUST GAS RECIRCULATION (EGR)

Unlike most technological improvements in fuel and ignition design which improve engine efficiency, the EGR exists to limit, not enhance combustion efficiency under certain operating conditions. EGR introduces a limited amount of inert, burned exhaust gas to the combustion chamber. This dilutes the air-fuel mixture, and limits combustion. The overall result is lower peak combustion temperatures and reduced NO_x emissions.

NO_x formation only occurs when combustion temperatures are high and the air-fuel mixture is lean. EGR flow is only allowed during cruise or moderate acceleration when the possibility of NO_x formation is greatest, and is prevented under low combustion temperature or rich operating conditions.

If the exhaust gas enters the cylinder during idle or at full throttle acceleration, performance suffers appreciably. In fact, EGR flow at idle may stall the engine. Therefore, controlling EGR opening time, as well as the amount of exhaust gas being recirculated, are critical for eliminating NO_x emissions and maintaining good driveability and satisfactory engine performance.

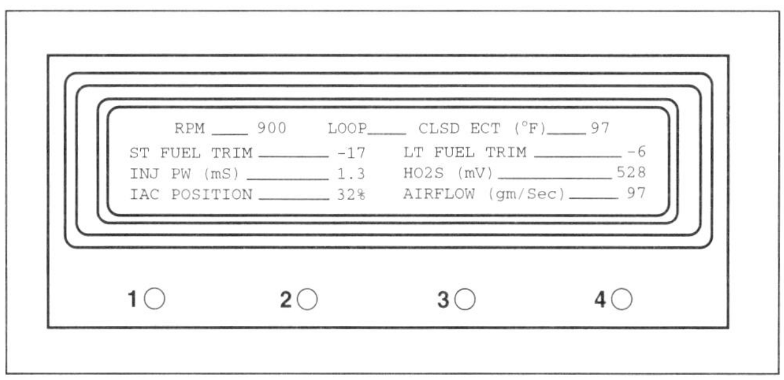

Figure 5-3 PCV problems cause a rich exhaust, the PCM responds to a high O2S signal by reducing fuel and increasing airflow.

EGR systems all operate on the same basic principles. Most use a vacuum diaphragm valve to meter exhaust gas flow into the intake charge, figure 5-4. Some systems are electronic and regulate EGR flow with pulse-width modulated solenoids.

On the composite vehicle, EGR flow is allowed when vacuum is applied to the diaphragm valve. The PCM controls the amount of flow by regulating the vacuum signal to the valve with a pulse-width modulated solenoid. EGR is allowed anytime the engine coolant temperature exceeds 150°F (66°C), and the throttle is neither wide open or closed.

EGR Malfunction Symptoms

An EGR system that has failed, or is malfunctioning, can cause:

- Poor idle quality
- Stalling
- Rough cold running and poor engine warmup
- Stumble, detonation or ping on acceleration
- Surge at cruise
- Poor high-speed performance
- Low engine vacuum.

EGR system malfunctions can be broken into two distinct categories:

- EGR flow restrictions — which fail to introduce enough exhaust gas to the combustion chamber to reduce NO_x
- Excess EGR flow — which introduces exhaust gas into the combustion chamber at the wrong time.

A quick way to check EGR valve operation on some systems is to manually open the EGR valve with the engine running at idle. The engine will stall if EGR is

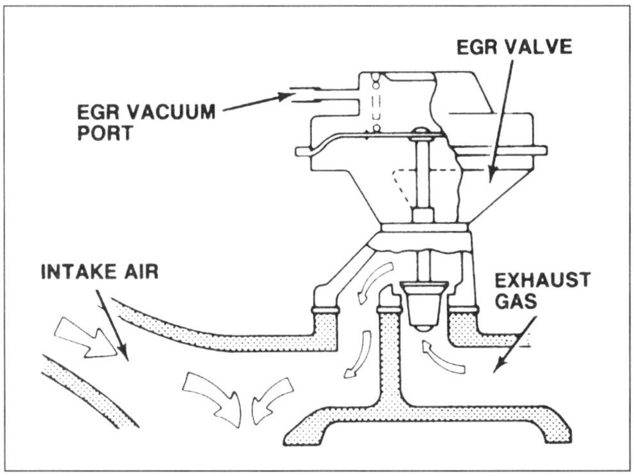

Figure 5-4 Typical EGR valve operation

flowing. However, this check is inconclusive as a partially plugged system can flow enough exhaust to stall an engine at idle, but fail to deliver its rated flow under load. This is an important consideration for I/M 240 testing where the vehicle is operated under light-load cruise conditions.

The source of an EGR system failure can be located in the:

- EGR system passages
- Vacuum control system
- Electronic control system
- EGR valve itself.

EGR system passage problems

Build up of exhaust deposits that form in the EGR passages, and on the valve itself, can prevent the valve from fully seating. The result is constant EGR flow. Monitoring data stream parameters may lead you to this conclusion, but the only way to verify it is to remove and inspect the valve assembly.

While many manufacturers recommend that the EGR be serviced at regular intervals to clean away the normal accumulation of carbon, most symptoms are simply ignored until the system stops working. Always clear any deposits or debris from the EGR passages and from the valve whenever it is removed.

Failures due to a flow restriction cause an increase in NO_x emissions. Since driveability symptoms are barely noticeable, this type of failure is rarely the cause of a customer complaint. Severe cases can result in detonation or pinging, and also affect fuel economy by preventing the control system from entering closed loop.

Suspect EGR flow restrictions when a vehicle fails an emission test due to high NO_x readings and all other exhaust gas levels are within specification.

Vacuum signal problems

A weak or missing vacuum signal to the EGR valve may keep it from opening. A constant vacuum signal will hold the EGR valve open all of the time. Begin by inspecting for missing or improperly routed vacuum hoses. Then, use a gauge to check for vacuum at the valve.

Some systems route vacuum to the EGR valve through a vacuum amplifier. A faulty amplifier can apply vacuum to the EGR valve at all times, regardless of throttle position, or fail to direct vacuum to the valve when needed. Verify that the correct amount vacuum is available at the EGR valve.

To provide good driveability, the system must prevent EGR flow during engine warm up following a cold start. Many systems accomplish this by routing the vacuum signal through a coolant controlled thermostatic vacuum switch, figure 5-5. A failed coolant vacuum control switch that prevents vacuum flow results in high NO_x emissions. A switch that allows the EGR valve to open on a cold engine can result in hesitation, poor response, and a lack of performance.

In addition, some systems only open the EGR valve in response to a combination of exhaust backpressure and vacuum. Even with a good vacuum signal, the low backpressure created by some "free flow", aftermarket exhaust systems can prevent EGR valve opening.

Electronic control problems

Late-model engine management systems regulate EGR valve opening with a PCM controlled solenoid. Most use a solenoid vacuum switch to activate a diaphragm, the PCM toggles the switch in response to various inputs. When testing these systems for flow, check the PCM signal, the solenoid, and the integrity of the vacuum system to and from the solenoid before condemning the EGR valve. Both digital, on/off solenoids, and pulse-width modulated solenoids are used.

When the PCM energizes an EGR solenoid, manifold vacuum is supplied to the diaphragm of the EGR valve, causing the valve to open. When the solenoid is not energized, the vacuum is blocked, and remaining vacuum is vented off allowing the valve to close, figure 5-6. The EGR valve should not open at idle or wide open throttle. Available data stream parameters vary by system, EGR valve information may show up as:

- Percent of EGR flow
- EGR duty cycle
- EGR on or off

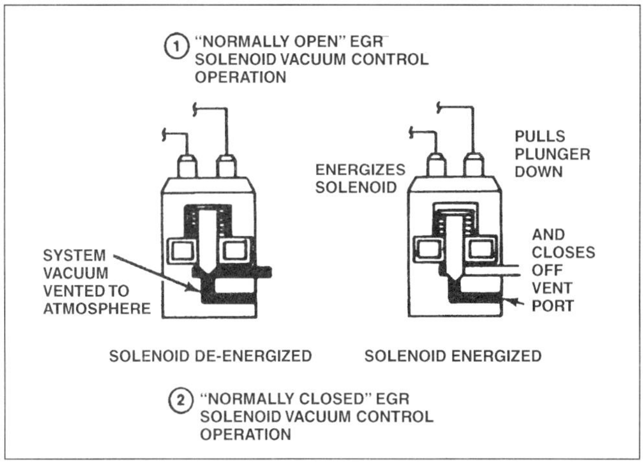

Figure 5-6 Solenoid control of the EGR valve vacuum signal.

The composite vehicle data stream shows solenoid position as a percentage. A 0 percent reading shows the valve is completely closed, and a 100 percent indicates the valve is fully open. Since this is a typical 12-volt circuit, solenoid signal voltage should be about 5.9 to 6.1 volts with a 50 percent data stream reading. Typically, a DTC is set when EGR flow is out of range, or flow is present at the wrong time. EGR failure on the composite vehicle will set DTC P0400, the only EGR code available.

EGR valve problems

EGR valves are designed to close when they fail. If the valve cannot open, EGR flow is prevented and NO_x emissions will be high. Occasionally a valve sticks in the open position, and carbon build up is often the cause. A small piece of carbon that lodges between the EGR pintle and seat, can hold the valve open at all times.

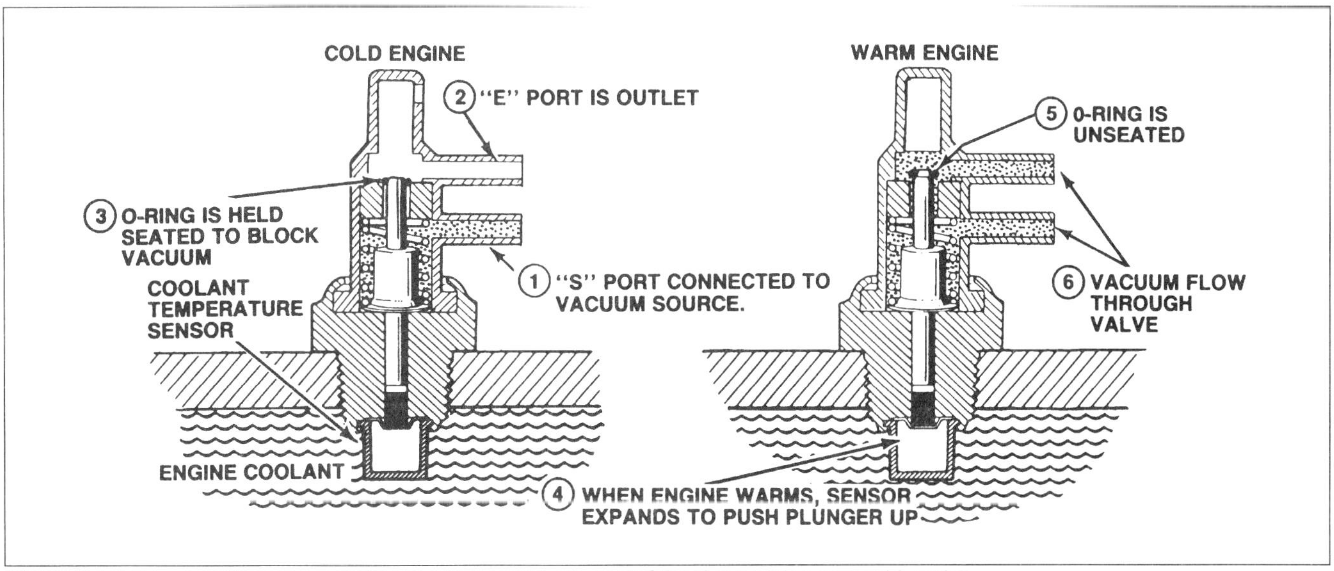

Figure 5-5 A thermostatic vacuum switch may be used to prevent EGR valve opening on a cold engine.

Eliminate the possibility of control system problems before condemning the EGR valve. While the EGR valve itself is a fairly simple mechanism, the systems which operate it can be very complex. Always refer to manufacturers specification and system descriptions before attempting to diagnose and repair an EGR system. General procedures for checking vacuum valve operation are in Chapter 4 of this book, and Chapter 2 includes procedures for checking solenoid valve operation.

EGR Driveability Problems and Emission Failures

The following are examples of some of the driveability problems and emission failures that can occur with an EGR system malfunction:

- Vacuum bleeding past the solenoid and partially opening the valve at idle can cause exhaust gas to displace the fresh air and fuel entering the cylinders. This results in incomplete combustion that causes high hydrocarbon emissions and possible engine stalling.
- Excessive EGR at cruise will cause power loss, surging, hesitation, and can also increase hydrocarbon emissions.
- No EGR flow, or flow that is below specifications at cruise, promotes engine detonation and results in excessive NO_x emissions.

IGNITION TIMING CONTROL

Precise spark timing control during all engine operating conditions is critical for maintaining acceptable exhaust emission levels. Therefore, the timing control system must be considered when diagnosing emission control problems.

With PCM controlled spark-timing, manufacturers generally provide specifications for checking base timing, as well as timing advance. You may have to disconnect the engine coolant sensor lead wire connector, figure 5-7, jumper across two terminals of the diagnostic connector, figure 5-8, or perform some other task in order to put the PCM in diagnostic mode. Base timing, 10° BTDC, is the only specification provided for the composite vehicle. Disable the advance by installing a jump wire between pin 3 and pin 4 in the data link connector to check base timing. Composite vehicle timing is adjustable.

To obtain optimum driveability, emissions levels, and fuel economy, ignition timing must vary in response to engine speed and load. The control system advances or retards timing as follows:

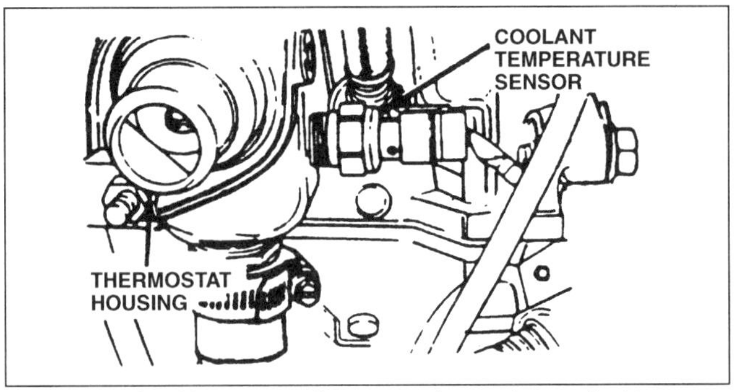

Figure 5-7 On this system, disconnect the engine coolant temperature sensor to check ignition timing.

- Timing must advance as engine speed increases, and retard as engine speed decreases.
- Timing must retard as engine load increases, and advance as load decreases.

When ignition timing does not respond to engine speed and load, combustion efficiency suffers. This results in a loss of performance and an increase in exhaust emissions. Incorrect timing will be reflected on manifold vacuum gauge readings and exhaust gas analysis results.

Timing Control Driveability Problems and Emission Failures

Ignition takes place too early when timing is over advanced.

Piston speed slows down because the piston is still on its compression stroke during combustion. This causes high combustion pressure and inefficient burning, which leads to detonation and high exhaust emissions. With advanced timing, the manifold vacuum gauge needle will be steady and rise 2 to 3 in-Hg (7 to 10 kPa) as the engine is brought off idle, figure 5-9. In-

Figure 5-8 On this system, jumper across two terminals of the diagnostic connector to check ignition timing.

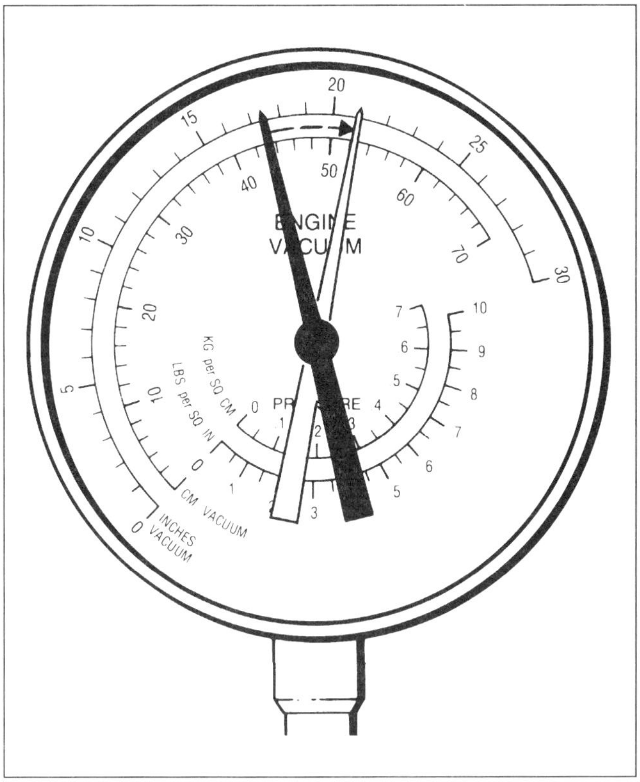

Figure 5-9 Advanced timing causes vacuum to rise slightly as the engine is brought off idle.

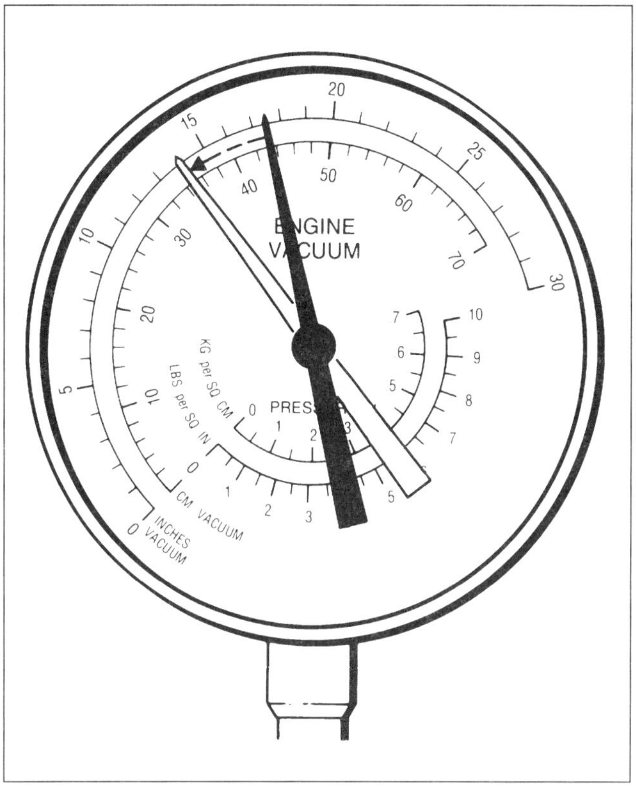

Figure 5-10 Retarded timing causes vacuum to drop slightly as the engine is brought off idle.

complete combustion causes a high hydrocarbon (HC) level with all other exhaust gases within specification.

With retarded timing, the piston is already well into its power stroke when ignition takes place. As a result, the effect of combustion on the piston is reduced. The fuel mixture burns slowly, but completely, and the engine suffers from a lack of power. On a manifold vacuum gauge, late, or retarded, timing will cause a steady needle to drop 2 to 3 in-Hg (7 to 10 kPa) as the engine is brought off idle, figure 5-10. Expect hydrocarbon (HC) readings to be below normal when all other exhaust gases are within specification.

IDLE AIR CONTROL (IAC)

As mentioned in Chapter 4, an idle air control motor regulates intake airflow, and has two basic functions:

- Regulate bypass airflow to maintain a desired idle speed.
- Regulate bypass airflow during closed throttle deceleration to maintain efficient combustion and good driveability.

Most manufacturers use a stepper motor to open and close the bypass air port. IAC failure can allow either too much or too little air into the engine, resulting in a too high or too low idle speed, respectively. An internal short can allow excess amperage into the PCM and damage the driver circuit. To avoid a new part failure, always check the resistance of the IAC windings, and all other actuators, when replacing a PCM.

IAC Driveability Problems and Emission Failures

A high, low, or erratic idle speed may indicate an IAC failure. Other possible symptoms include tip-in sag, hesitation, and stalling on deceleration.

Restricted bypass airflow due to deposit build up on the IAC valve and passages is a common problem with multiport injection systems. Expect IAC step counts or percent to be high as the valve tries to flow more air around the restriction. Idle speed and the IAC signal may be erratic.

The composite vehicle will set DTC P0505 if the PCM detects a malfunction in the idle speed control system. Specific information on what conditions set the code is not provided. Most systems will only set a DTC if the signal is absent, or the IAC motor is at its high or low adaptive limit and cannot maintain idle speed within the specified tolerance. Therefore, most IAC failures result in a no code driveability problem.

CATALYTIC CONVERTER SYSTEM

In chemistry, a catalyst is a substance which promotes a chemical reaction without entering into the reaction or being changed by that reaction. In a catalytic converter, a catalyzing agent chemically transforms harmful gases into harmless ones by promoting this type of chemical reaction.

Two types catalyst, oxidation and reduction, are used in converters. The type of chemical reaction that takes place marks the difference between the two:

- Oxidation catalysts add oxygen
- Reduction catalysts remove oxygen.

Either palladium or platinum can be used as the catalyst to promote a reaction in an oxidation converter. Oxidation effects two exhaust gases, hydrocarbons (HC) and carbon monoxide (CO). The addition of oxygen combines with HC and CO to form harmless carbon dioxide (CO_2) and water (H_2O).

Rhodium is used as the catalyst in a reduction converter. Reduction lowers the level of oxides of nitrogen (NO_x) in the exhaust. The rhodium strips oxygen from NO_x and releases nitrogen and oxygen in its place.

Newer converters combine both an oxidation and a reduction catalyst in a single housing, figure 5-11. Two converter designs, the dual bed and the three way, are in use.

Dual-bed Catalytic Converters

Oxidation occurs first with a dual-bed converter. As exhaust enters the reducing bed, oxygen is stripped from NO_x compounds and passes into the oxidizing bed as nitrogen and oxygen. Next, the reduction catalyst uses the left over oxygen from the oxidation process to convert HC and CO into CO_2 and water.

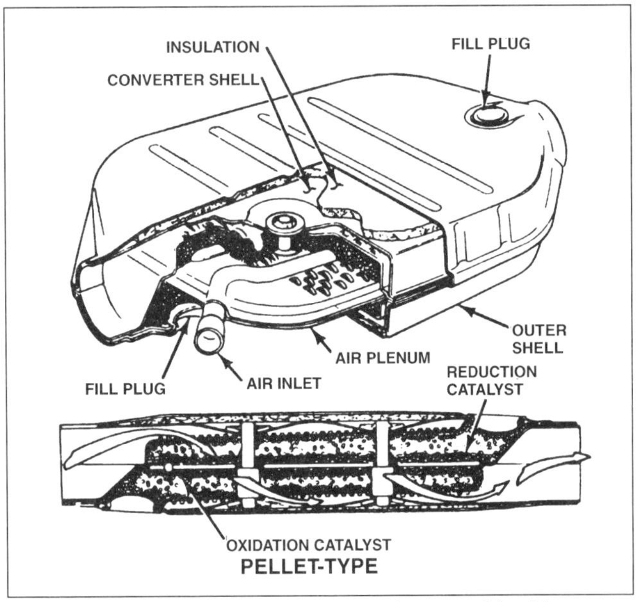

Figure 5-11 Oxidation and reduction catalysts combined in a common housing.

During normal operation, the reduction process does not release enough free oxygen for the oxidation catalyst to completely oxidize all of the HC and CO. For this reason, additional oxygen must be supplied to the oxidation bed of the converter. This task is performed by the secondary air system. Under certain operating conditions, secondary air is directed to the oxidation section of the converter to promote efficient operation and lower emissions.

Three-way Catalytic Converters

With a three-way converter, a hybrid catalyst that promotes both reduction and oxidation is used in combination with a second oxidation catalyst. The front chamber, or bed, of the converter contains the hybrid catalyst. So, the combined reduction and oxidation process takes place first. Then, the exhaust passes into the second chamber where remaining HC and CO are exposed to, and broken down by the second oxidizing catalyst, figure 5-12.

Like its dual-bed counterpart, a three-way catalytic converter requires the addition of secondary air under certain conditions. The demand for extra oxygen is highest following a cold startup. As the converter comes up to operating temperature, the need for additional air diminishes as catalyst efficiency increases.

Requirements for Efficient Converter Operation

Quickly reaching operating temperature is critical for efficient converter operation. A catalyst must reach about 500°F (260°C) before it can promote a chemical reaction. Internal temperature on an operating converter will be higher than exhaust gas inlet temperature. A fully effective converter operates in the 900 to 1600°F (482 to 871°C) temperature range. Note, this is still well below the 2500° degrees at which NO_x forms.

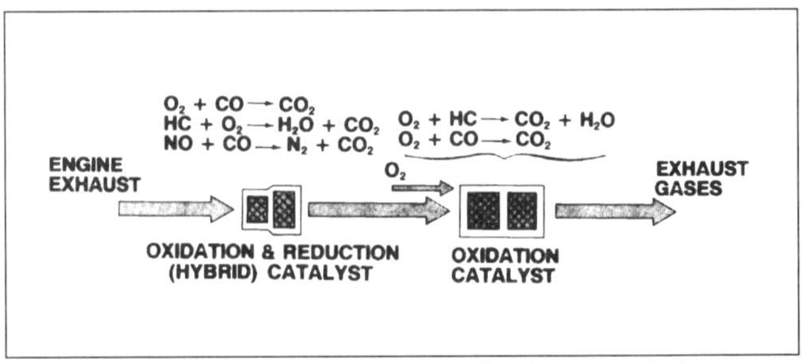

Figure 5-12 Hybrid catalysts oxidize and reduce gases, the exhaust then passes to a second oxidation catalyst.

Outlet temperatures at the catalyst can be up to 200°F (93°C) higher than inlet temperature. A catalyst whose outlet temperature is cooler than or equal to the inlet temperature is not working.

Continued exposure to an overly rich mixture will destroy a catalyst. As the catalyst attempts to break down the high levels of HC and CO from the rich mixture, internal temperature rises. Eventually, the converter overheats and the catalyst chemically breaks down rendering it ineffective. In an effort to prevent overheating of the exhaust during prolonged idling conditions, downstream air is eliminated by some manufacturers after the engine has idled for a specified period of time.

Lean operating conditions also effect catalytic converter efficiency. A lean mixture contains very little CO and HC which are easily oxidized. However, lean running conditions also promote high NO_x formation, and the reducing catalyst is not as effective as its oxidizing counterpart. A catalyst is most efficient when the engine is operating at a stoichiometric air-fuel ratio.

In addition to a correct air-fuel ratio, catalytic converter efficiency also depends on proper secondary air routing and EGR system operation. To minimize oxidation requirements, combustion temperature must remain below 2500°F (1370°C) to prevent NO_x formation. Ignition and cooling system problems can effect converter operation as well.

Converter Driveability Problems and Emission Failures

Except in the case of a complete or extreme converter failure, symptoms of a converter malfunction are slight and generally go unnoticed by the customer.

A typical complaint would be a strong "rotten egg" odor. The odor results from the formation of hydrogen sulfide H_2S, a toxic byproduct of converter action. Hydrogen sulfide naturally occurs in small quantities, and formation is heaviest during engine warm up and deceleration. A strong odor from an engine running at normal operating temperature is generally an indication of a rich air-fuel mixture. Continued operation can lead to converter failure.

Converter failure often causes the catalyst to break down and restrict the flow of exhaust gases. The increase in backpressure can cause complaints of a lack of power, especially when under load. See Chapter 4 for details on diagnosing a restricted exhaust.

Exhaust gas analysis is the most effective method of determining catalytic converter efficiency. Preconditioning the engine for emission testing as outlined in Chapter 1 of this book. Suspect a faulty catalytic converter when all of the following emission readings are present:

- CO exceeds 0.3 percent
- O_2 exceeds 0.4 percent
- HC is greater than 120 ppm, but less than 400 ppm

These guidelines are simply a rule of thumb, however they will be accurate about 90 to 95 percent of the time.

SECONDARY AIR INJECTION (AIR)

Originally, secondary air was introduced into the exhaust manifold where it would combine with hot exhaust gases leaving the cylinders. The additional air promotes oxidation to effectively lower the HC and CO content in the exhaust. The need for secondary air injection to reduce HC and CO emissions on newer vehicles has been reduced due to the precise air-fuel metering capabilities of electronic control systems. In fact, the ASE composite vehicle does not use secondary air injection.

However, as previously mentioned, some catalytic converters do require secondary air to operate efficiently. On late-model systems, secondary air injection switches between several different locations depending on operating conditions, figure 5-13. Secondary air may be introduced:

- Near each exhaust port
- Into the exhaust manifold outlet
- Into the exhaust pipe ahead of the converter (upstream)
- Directly into the converter (downstream).

All secondary air systems are relatively simple, and two basic designs are used:

- Air pump systems
- Pulse air systems.

Air Pump

An air pump is belt driven off the crankshaft. The pump pulls in fresh air and delivers slightly pressurized air to the exhaust. Most pumps use a pressure relief valve, figure 5-14, which opens at high engine speed. Pumps without a pressure relief valve use a pressure relief valve, built into the control valve.

Systems used with a three-way converter generally have a switching device to direct airflow. Depending upon operating conditions, secondary airflow is either directed to the exhaust manifold, the converter, or vented to atmosphere. Air injected into the exhaust upstream from the converter raises the exhaust gas tem-

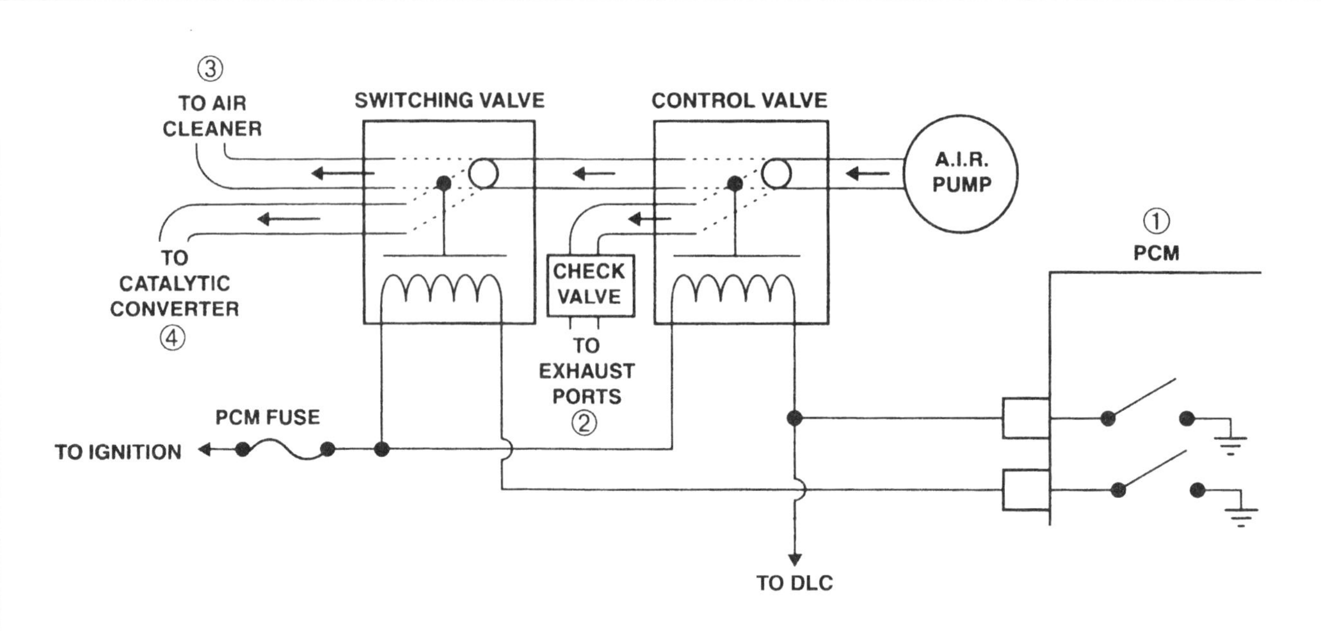

Figure 5-13 On electronic systems, the PCM (1) directs secondary airflow to either the exhaust (2), atmosphere (2), or converter (3).

perature to help the converter reach operating temperature quickly. Air injected directly into the converter provides the additional oxygen needed to oxidize HC and CO completely.

Antibackfire Valves

During deceleration, high manifold vacuum creates a rich air-fuel mixture. Secondary air flowing into the exhaust manifold at this time will combine with unburned fuel and result in a backfire. To prevent backfire, secondary air must be shut off, or the mixture leaned, during deceleration. Two types of diaphragm-operated valves, the gulp valve and the diverter valve, are used to redirect the airflow.

Gulp valve

When manifold vacuum is increased to a gulp valve diaphragm, secondary air flows to the intake and the exhaust manifolds, figure 5-15. This leans out the air-fuel mixture to prevent backfiring. Balance restrictions, or bleed holes, in the valve assembly equalize pressure on both sides of the diaphragm after a few seconds to close the valve.

Any sudden increase in vacuum will operate the gulp valve. A gulp valve that is stuck open can cause hard starting and a rough idle. Momentarily closing the throttle at high speeds while manifold vacuum is low can delay valve operation, and cause a backfire or hesitation.

Diverter valve

With a diverter valve, secondary air circulates through

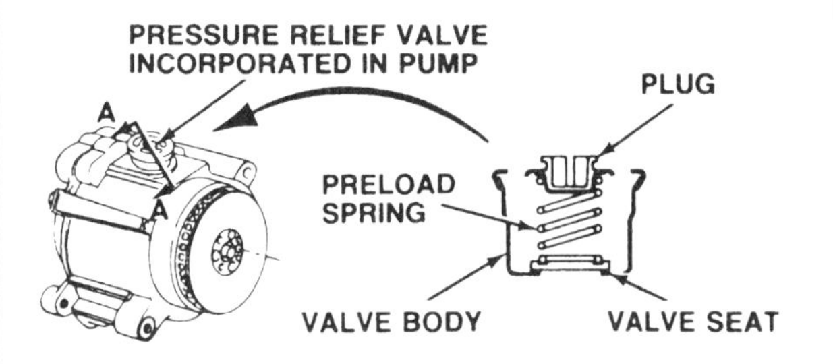

Figure 5-14 A relief valve vents pressure at high pump speeds

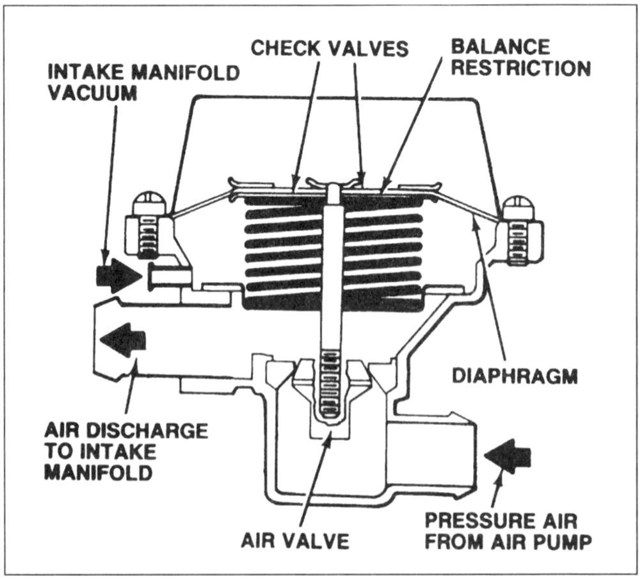

Figure 5-15 A gulp valve flows secondary air to the intake when vacuum is applied.

the valve continuously. During deceleration, manifold vacuum operates the valve to divert airflow directly to the atmosphere, rather than to the intake manifold.

Diverter valves may be used along with vacuum solenoids, vacuum differential valves, vacuum vent valves, and idle valves to fine-tune air injection, figure 5-16. A diverter valves has no effect on the air-fuel mixture in the intake manifold.

Exhaust Check Valves

All air injection systems use one or more one-way check valves, figure 5-17, to protect components from reverse exhaust flow. A check valve contains a spring-type metallic disc or reed that closes the air passage under exhaust backpressure.

Air Management Valves

Air management valve is a term used to describe the large variety of assemblies that combine the divert and switching functions into one unit. The divert, or air control, section protects the converter during open-throttle operation and high temperatures. The air switching section sends air to the exhaust ports during open-loop operation. When the control system is operating in closed loop, the valve switches airflow to the converter.

A single valve or a combination of several valves may be used to redirect fresh air:

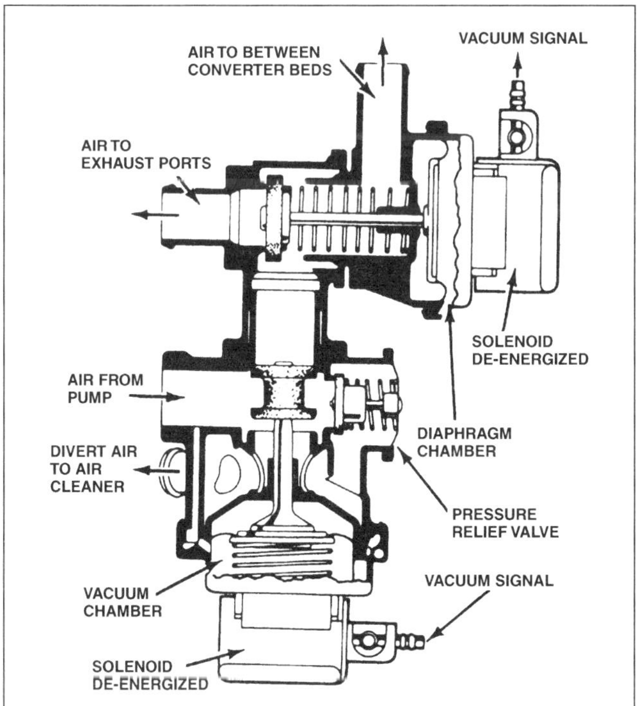

Figure 5-16 An electronic diverter valve used for routing secondary air.

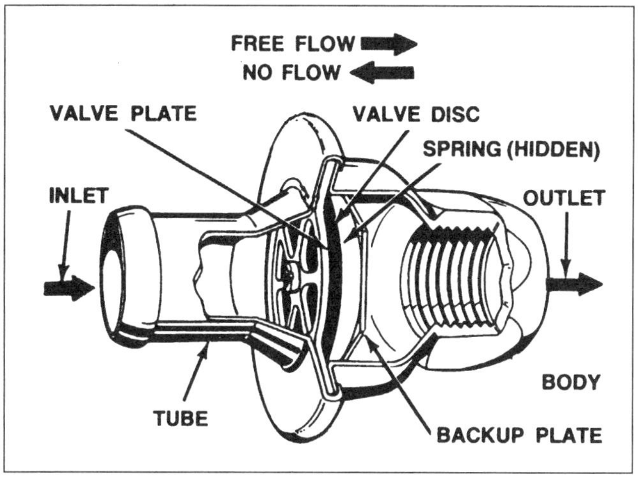

Figure 5-17 Typical secondary air check valve.

- To the exhaust ports or manifold when the engine is first started to help bring the O2S and converter to operating temperature, and to help control HC and CO emissions.
- Downstream to help the oxidation converter in warm-engine control of HC and CO.
- Away from the exhaust during deceleration to prevent backfiring and protect the converter from an overheat condition.
- Away from the exhaust during wide-open throttle or heavy load conditions to protect the converter from overheating.
- Away from the exhaust if the engine overheats or a component malfunctions. This also protects the converter from damage caused by overheating.

Pulse Air Injection

Pulse air injection uses the low pressure pulses that naturally occur in a running engine to draw fresh air into the exhaust. This eliminates the need of a belt-driven pump, reducing weight and parasitic loss.

Pulse air, or aspirator, valves are similar in design to a check valve. Each time an exhaust valve closes, manifold pressure momentarily drops below atmospheric pressure. When pressure is low, the pulse valve opens to admit fresh air, figure 5-18. When exhaust pressure is greater than atmospheric pressure, the valve closes. Late-model pulse air systems are switched on and off, depending on operating conditions, by a PCM controlled switch.

AIR Driveability Problems and Emission Failures

Driveability complaints that relate to the secondary air system include excessive noise caused by a defective pump or misdirected air, and backfire during decelera-

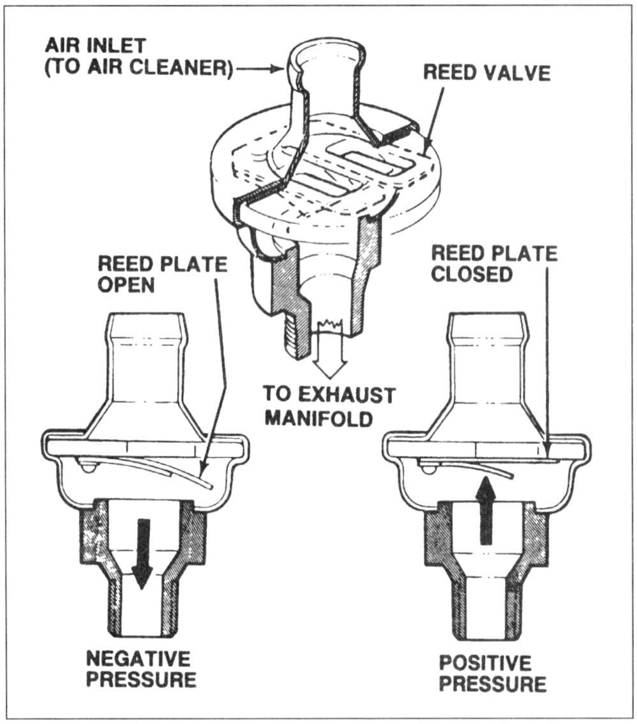

Figure 5-18 Pulse air valves open to let in air when there is negative pressure in the exhaust manifold.

tion caused by a valve malfunction. Failures generally do not set a diagnostic trouble code. However, since misdirected secondary air can effect the air-fuel mixture, failure may cause fuel correction codes to set.

Excess airflow during an emissions test will cause HC, CO, and CO_2 readings to drop, and O_2 readings to increase. Lack of secondary airflow causes a richer condition, HC, CO, and CO_2 rise, while the O_2 level drops.

INTAKE AIR TEMPERATURE (IAT) CONTROL

As mentioned in Chapter 4, a temperature sensing device is used to direct either cold or preheated air to the intake manifold. The primary purpose of regulating intake air temperature is to reduce hydrocarbon (HC) and carbon monoxide (CO) emissions during engine warmup. Malfunctions can cause emission test failure.

IAT Driveability Problems and Emission Failures

Driveability is effected when the intake air temperature control system is not working correctly. If intake air temperature is too low, engine performance is erratic. This causes engine stalling and poor throttle response on acceleration. High intake air temperatures cause power loss and possibly detonation. Complaints include an engine that seems to be running out of fuel, and rough, or erratic automatic transmission shifting. Either type of failure can cause exhaust emissions to increase.

EARLY FUEL EVAPORATION (EFE) CONTROL

Early fuel evaporation control applies to the manifold heat control valves discussed in Chapter 4. In addition, some engines use an electric grid installed in the intake air stream to preheat the intake charge, figure 5-19. Check for proper function of the heating element when cold driveability problems exist.

EFE Driveability Problems and Emission Failures

Failure of the EFE system can cause a number of driveability complaints, and increase exhaust emissions. High intake air temperature can cause increased fuel consumption and poor engine performance. Low intake air temperature can cause poor acceleration, a lack of power, and poor high-speed performance.

EVAPORATIVE EMISSION (EVAP) CONTROL

Vapor from the evaporation of raw fuel in the gas tank and other fuel system components is a major source of HC emissions. It has been estimated that 20 percent of all vehicle hydrocarbon emissions came from unregulated fuel evaporation. The evaporative emission (EVAP) system traps, stores, and then burns evaporated fuel to reduce these passive emissions.

Fuel evaporation controls are fairly simple, figure 5-20, but often ignored by service technicians unless there is a major driveability problem. However, EVAP function is very important to overall emission control. As part of new I/M 240 Enhanced Tests, the integrity of the EVAP system is tested.

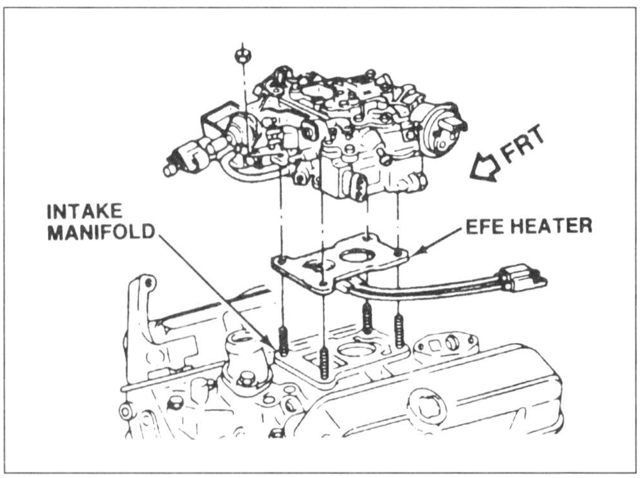

Figure 5-19 An electric EFE heating grid.

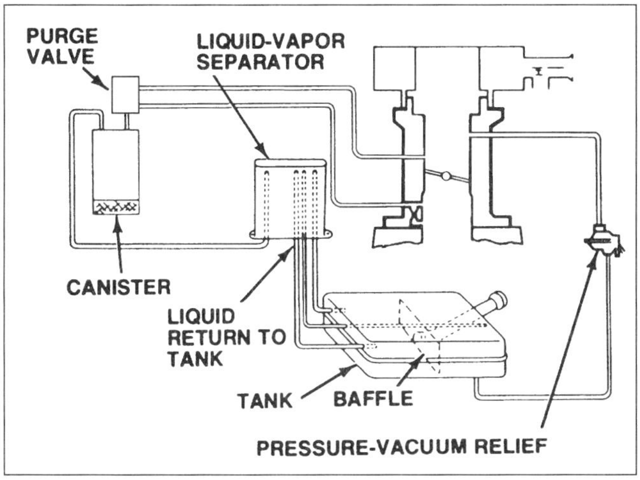

Figure 5-20 Typical evaporative emission control system.

While EVAP systems vary from manufacturer to manufacturer, most use common components, such as:

- Sealed fuel filler cap
- Fuel tank that allows for expansion
- Vapor storage canister
- Liquid separator
- Purge control device.

Most filler caps have a pop-off valve which vents pressure over about 1 psi. A reed valve allows fresh air in to equalize tank and atmospheric pressures when atmospheric pressure is greater. All fuel tanks have a dead space which allows for normal expansion of fuel, and many incorporate a limiter to prevent overfilling.

Vapor storage canisters are simply containers filled with activated charcoal which store fuel vapor until it can be drawn into the engine and burned. The canister is designed to store vapor only, the bottom of the canister is open to atmospheric pressure. Liquid separators prevent liquid fuel from reaching the canister by allowing only vapor to pass.

EVAP Canister Purge Control

Vehicle manufacturers have used a wide variety of purge control devices over the years. Early canisters were simply connected to manifold vacuum. Manifold vacuum draws off the vapor to purge the canister as soon as the engine starts. Later systems use ported vacuum. This way vapor is only drawn in when engine speed is above idle. At this point, the engine is better able to burn the additional fuel. Some systems use a temperature control valve that allows purging only when the engine is warm.

On current vehicles, canister purging is a PCM controlled function. Operation is similar to that of EGR solenoids mentioned earlier. The PCM controlled EVAP solenoid switches a vacuum signal to allow or prevent purging based on engine speed, temperature, and other inputs, figure 5-21. The Composite vehicle's system allows purge whenever the engine coolant temperature is over 150°F (66°C) and the throttle is open. When energized, engine vacuum purges fuel vapors (HC) from the charcoal canister.

Purge control problems

Problems with the purge system often result in a rich air-fuel mixture due to the high hydrocarbon content of the vapors. Most control systems set a diagnostic trouble code (DTC), such as P0440 used on the composite vehicle, when an EVAP malfunction is detected. Failure can also set related fuel correction codes and effect serial data as well. The integrity of the EVAP system must be monitored by the on-board control system to comply with OBD-II regulations. When available, canister purge information usually shows up in a data stream as: EVAP on/off.

I/M 240 Testing

I/M 240 requires two tests, vacuum purge and pressure, on the EVAP system. We explain each briefly here.

Vacuum purge test

This test determines if the purge control is actually drawing vapor from the canister. A flow meter, which measures both instantaneous and cumulative vapor flow, connects with a tee to the purge line between the canister and the engine. The flow meter remains in place during the entire I/M 240 drive cycle. To pass, the EVAP system must purge a cumulative volume of at least 1 liter of vapor during the 240 second test. This is a small volume compared to the total amount of vapor purged by a normally operating system, and may represent only 1/25, or 4 percent, of normal flow rates.

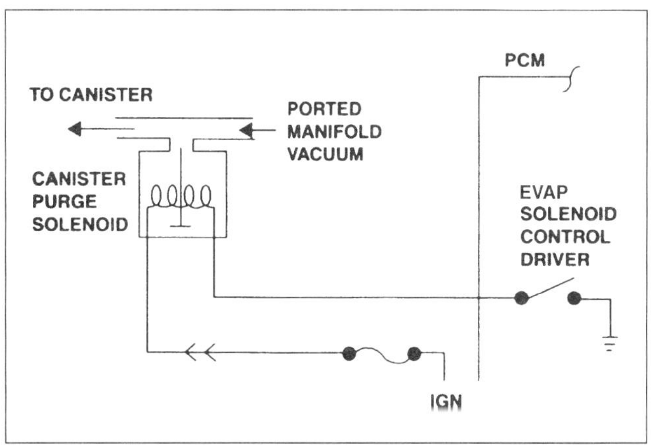

Figure 5-21 Late-model systems use a PCM output signal to regulate canister purging.

Evaporative pressure test

This test checks the system for leaks which can allow vapor to escape into the atmosphere. The test is conducted with the vehicle at rest and the engine not running.

During the test, a pressure gauge connects to the vapor vent line between the canister and the fuel tank, and the system is pressurized with nitrogen gas. Nitrogen is introduced to charge the system to a pressure of 14 inches of water (0.5 psi). The EVAP system must hold pressure above 8 inches of water for two minutes in order to pass.

Failure can often be traced to a cracked or disconnected vapor line. Look for signs of modification or tampering. A replacement fuel filler cap that does not meet original specifications can cause the system to fail a pressure test.

EVAP Driveability Problems and Emission Failures

Canister purging at idle causes a rich mixture that can result in a rough idle and increases tailpipe emissions. Customer complaints of strong fuel odors often result from broken vent lines or a fuel saturated canister. This type of failure has no effect on a standard emissions test, it will show up during an I/M 240 pressure test. Most failures cause an increase in passive emissions only. The following are examples of some common EVAP system problems:

- A faulty EVAP system that has pressure leaks will cause hydrocarbons to be admitted to the atmosphere.
- An EVAP solenoid that does not open during PCM commands will cause excessively high pressures to build up in the fuel systems and can cause filler neck and other fuel system seals to leak.
- An EVAP solenoid with low resistance will cause excessive current flow through the PCM driver circuit causing possible damage to the driver circuit.

1. An engine idles roughly and stalls on light acceleration. When the vacuum line is disconnected from the EGR valve, the problem disappears. This probably means:
   a. The EGR valve is stuck closed
   b. The EGR valve is getting a weak vacuum signal
   c. The EGR valve diaphragm spring is weak or broken
   d. The EGR valve malfunction is causing the engine to produce excessive NO_x.

   Explanation: The reason for this condition is that the EGR valve is not sealing properly and is coming on too soon when it receives vacuum. When the spring is weak or broken, the EGR valve may not seat properly, so c is the correct answer. As soon as it gets vacuum, during light acceleration, it will cause the engine to run roughly and may even stall but will not cause excessive NO_x.

2. During inspection of an air injection system, a brittle, burned hose is found on the upstream side of a check valve. This may be due to:
   a. The hose being positioned too close to the engine
   b. The air pump developing excessive air pressure
   c. The injection nozzles being plugged and allowing backpressure to build up in the system
   d. The check valve allowing exhaust gas to leak back into the air injection system

   Explanation: The whole purpose of the air injection check valve is to prevent gases from leaking back into the air system. When this happens, hoses and even the air filter itself can become burnt, so the answer is d.

3. A car with an air pump emission control system backfires when decelerating. Which of these should the technician check?
   a. The operation of the exhaust manifold check valve
   b. The output pressure of the air pump
   c. The operation of the diverter or gulp valve
   d. The air manifolds for leakage

   Explanation: The purpose of the diverter or gulp valve is to prevent air from entering the engine via the exhaust valve ports on deceleration. This would cause the engine to backfire. Therefore, c is the best answer.

4. An evaporative system pressure test is being performed at the canister purge line using a manometer pressurized to 14 inches of water. A hissing sound can be heard near the gas cap and high levels of HC are being measured when the exhaust analyzer probe is close to the gas cap or filler neck.

   Technician A says this could be the result of a wrong or defective gas cap.

   Technician B says this could be caused by a damaged filler neck that does not allow the cap to seal well.

   Who is right?
   a. a only
   b. b only
   c. Both a and b
   d. Neither a nor b

   Explanation: Either the gas cap or the filler neck could be the cause since there should not be any leakage of HC at this pressure. Both Technicians are correct.

5. The resistance of the canister purge valve solenoid is being tested on the composite vehicle and measures between 3 and 4 ohms.

   Technician A says this solenoid is faulty and must be replaced.

   Technician B says the PCM may have become damaged due to excessive current caused by the solenoid.

   Who is right?
   a. a only
   b. b only
   c. Both a and b
   d. Neither a nor b

   Explanation: A 3 to 4 ohms reading on a canister purge solenoid is almost a direct short and will draw 10 to 15 times its normal current, which could damage most PCM's in a short period of time. Both a and b are possible, so c is the correct answer.

6. During engine warm-up with the ECT temperature at 140°F and the throttle partially open, what voltage should the technician expect to see at pin #49 (EVAP solenoid) on the composite vehicle?

   a. 0 volts

   b. 0.05 to 0.75 volts

   c. 5 volts

   d. 12 to 14 volts

   Explanation: The EVAP solenoid will not be energized until after 150 degrees Fahrenheit. Therefore, pin #49 will read between 12 to 14 volts.

7. Technician A says that air is injected into the catalyst to lower catalyst temperatures and limit NO_x formation.

   Technician B says that air injection to the oxidizing bed of a dual bed catalyst helps oxidize HC and CO converting them to CO_2 and water.

   Who is right?

   a. Technician A

   b. Technician B

   c. Both A and B

   d. Neither A nor B

   Explanation: When the engine is warm and running in closed loop, air is pumped into the catalyst on some cars to provide additional oxygen for the oxidation process which converts CO and HC to CO_2 and water. Technician B is correct. Technician A is incorrect because NO_x emissions are converted to nitrogen and oxygen in another part of the catalyst. This process strips the NO_x of its oxygen and is called reduction. Downstream air is not pumped to the reduction part of the catalyst. In the oxidizing part of the catalyst, it actually increases catalyst temperature by promoting the chemical oxidation process.

8. During an I/M 240 evaporative pressure/purge system test the evaporative system is pressurized to 14.5 inches of water. After 2 minutes the pressure has dropped to 10 inches of water.

   Technician A says this could be caused by excessive leakage at the canister purge solenoid.

   Technician B says this is acceptable and no repairs are required.

   Who is right?

   a. A only

   b. B only

   c. Both A and B

   d. Neither A nor B

   Explanation: When the evaporative system is pressurized to 14.5 inches of water for a period of 2 minutes, the system is acceptable if a pressure of at least 8 inches is maintained, so Technician B is correct.

9. An exhaust gas temperature sensor is being tested on a customer's car. Which of the following is true about this sensor?

   a. It is a negative temperature coefficient thermistor

   b. It operates on a 5 volt reference

   c. A voltage of 2.5 would indicate that the EGR valve has been hot and working

   d. All of the above

   Explanation: Virtually all gas temperature sensors operate on a 5 volt reference. They are a negative temperature coefficient thermistor type of sensor. For the voltage to drop to 2.5 volts, it would indicate that the sensor was hot because of the operation of the EGR valve.

Chapter Six

I/M FAILURE DIAGNOSIS

To perform inspection and maintenance (I/M) failure diagnosis, you must understand the byproducts of combustion. This is true whether you perform typical no-load emission testing at idle or run fully loaded I/M 240 emission tests on a chassis dynamometer.

The previous chapters covered the parts and systems that affect engine driveability, performance, and exhaust emissions. This chapter covers the gases emitted from the tail pipe. We cover the makeup and relationships of the gases and explain the circumstances or problems that cause the gases to be excessive or within specifications.

THE BYPRODUCTS OF COMBUSTION

We can measure up to five gases when testing exhaust emissions:

- Carbon Monoxide (CO)
- Hydrocarbons (HC)
- Oxygen (O_2)
- Carbon Dioxide (CO_2)
- Oxides of Nitrogen (NO_x)

CO, HC, and NO_x are the object of emission controls. Hydrocarbons and carbon monoxide are probably familiar to you through typical emissions testing. With the advent of I/M 240 testing, we have to consider oxides of nitrogen as well. HC and CO are the direct result of incomplete combustion and NO_x is the product of combustion at very high temperatures.

CO_2 and O_2 are commonly measured as an indication of combustion efficiency. Neither is harmful. Carbon dioxide is a product of complete combustion and its chemical notation, CO_2, indicates that carbon and two complete molecules of oxygen have combined during combustion (unlike CO which has bonded with a single oxygen atom as a result of incomplete combustion).

If we have complete combustion, all of the C and O_2 will combine to form CO_2. However, since combustion is never 100 percent efficient, some CO and leftover oxygen pass through the combustion chamber and into the exhaust. As a result, you can use either gas as an indicator of a rich or lean running engine. However, CO is only effective as a rich indicator when the ratio of air-to-fuel in the mixture is about 14.4 to 1 or less. Oxygen content is low when the mixture is rich, and high when there is not enough fuel available during a lean condition. The oxygen sensor generates a voltage signal based on the O_2 content of the exhaust which is used as a rich-lean indicator by the control system.

NO_x emissions, a byproduct of very high combustion temperatures (2500°F and above), form when nitrogen and oxygen combine. Although these pollutants have always been present, and auto manufacturers have been installing components to minimize them for quite some time, NO_x emissions were not monitored during field testing and emissions certification. This will change with the implementation of I/M 240 testing. I/M 240 regulations require you to measure NO_x emissions and perform repairs and adjustments to reduce unacceptably high levels of NO_x emissions.

Four-gas analyzers measure the parts per million (ppm) of HC, the percentage of CO, the percentage of O_2, and the percentage of CO_2. Most properly tuned, computer-controlled vehicles will produce less than 50 ppm of HC, less than 0.5 percent CO, 1.0-2.0 percent O_2, and 13.8-16.0 percent CO_2, figure 6-1.

In this chapter, we look at each of these four ever-present gases individually, followed by a discussion of NO_x formation which only occurs under special cir-

TYPICAL READINGS AT IDLE

YEAR	HC	CO	CO_2	O_2
65	700	4.0	12.0	0.5
70	150	2.0	13.0	0.5
75	60	0.60	14.0	1.0
80	45	0.50	14.0	1.5
85 & up	30	0.30	16.0	2.0

Figure 6-1. Typical tail pipe readings during an emissions test. The air-injection system is disabled for more accurate readings.

cumstances. Then, we establish the relationship between these gases and explain how test results are used to evaluate the integrity of the emission control system of a vehicle. In addition to the discussion that follows, refer to the table in figure 6-2 for problems that may cause abnormal readings.

Carbon Monoxide

By chemical composition, gasoline is a complex structure built on a compound hydrocarbon base. Hydrocarbons are simply molecules of hydrogen and carbon that are chemically bonded together. During the combustion process, chemical bonds break down to release free carbon which combines, or bonds to, free oxygen. This bonding results in the formation of either CO or CO_2, or both. CO is a highly-toxic, unwanted by-product of incomplete hydrocarbon combustion and is a major pollutant. On the other hand, CO_2 is a relatively harmless substance that forms during complete combustion and as a result of catalytic conversion.

Some CO will always be present in the exhaust, at least as long as engines continue to be less than 100 percent efficient. While CO is a product of incomplete combustion, it is not the best indicator of combustion efficiency.

A rich-lean indicator

You can think of CO as a rich-lean indicator when considered with the presence of oxygen. In fact, as an engine approaches the correct air fuel ratio, CO and O_2 should be evenly balanced, indicating that the air-fuel mixture is neither too rich nor too lean, but balanced.

The CO and O_2 relationship is a delicate balance that depends on the correct air-fuel mixture entering the engine, and an efficient burning of the fuel charge. CO is created during the combustion process due to a lack of oxygen. If we alter the amount of either air or fuel available for combustion, the amount of CO formation changes as well. Anything that *decreases* the amount of air getting to the combustion chamber or *increases* the flow of fuel to the combustion chamber will increase CO content in the exhaust.

High CO indicates a rich mixture

For CO to increase, either air *decreases*, fuel *increases*, or both. The relationships are inverse. Think about

	EMISSION	IDLE	IDLE OFF	CRUISE 1800-2000	AIR/FUEL RATIO	CAUSES POSSIBLE	SYMPTOMS RELATED
1	CO	3%	3%	3%	Rich AFR Below 10:1	Vacuum leak to map sensor Fuel injectors leaking Bad power valve (carburetor) Excessive fuel rail pressure Vacuum diaphragm bad Faulty Sensor Fuel level too high	Black smoke or sulphur odor Poor fuel economy Surge/hesitation Stalling Engine not preconditioned Engine in open loop
	HC	250 ppm	280 ppm	300 ppm			
	CO_2	7-9%	7-9%	7-9%			
	O_2	.2%	.2%	.2%			
2	CO	1.5%	1.5%	1.0%	Rich AFR at low speed 10-12:1	Engine oil diluted with fuel Carburetor idle speed (low) Cold engine Idle mixture set too rich Choke stuck shut PCV valve defective Fuel injectors leaking	Poor fuel economy Sooty spark/black smoke Rough idle/surge hesitation Vapor canister purge valve bad Vapor canister saturated
	HC	150 ppm	150 ppm	100 ppm			
	CO_2	7-9%	7-9%	11-13%			
	O_2	0.5%	0.5%	1.0%			
3	CO	0.5%	0.5%	0.1%	Lean AFR Over 16:1	Check ignition primary/secondary Vacuum leak (carb base, intake runner) EGR valve, vacuum lines Carburetor mixture too lean Poor cylinder sealing (rings/head) Fuel injectors restricted Improper timing Exhaust valve leak	Rough idle Misfire - high speed Detonation (cruise 2000 rpm) Surging Idle hunting (computer) Overheating NOTE: Excessive misfire and overheating can damage converter.
	HC	200 ppm	200 ppm	250 ppm			
	CO_2	7-9%	7-9%	7-9%			
	O_2	4-5%	4-5%	4-5%			
4	CO	2.5%	1.0%	0.8%	Lean AFR at High Speed Over 16:1	Air cleaner heater door closed Internal carburetor problem (float stuck, wrong jet, metering rod stuck) Fuel injectors restricted Fuel pump pressure low	Rough idle Misfire Surging/hesitation
	HC	100 ppm	80 ppm	50 ppm			
	CO_2	7-9%	7-9%	7-9%			
	O_2	2-3%	2-3%	2-3%			
5	CO	0.3%	0.3%	0.3%	AFR 13:1 - 15:1	Engine not preconditioned Air management system not disabled Converter not warmed	None No driveability symptoms
	HC	100 ppm	80 ppm	50 ppm			
	CO_2	10-12%	10-12%	10-12%			
	O_2	2.5%	2.5%	2.5%			

Figure 6-2. These 4-gas readings can help with emission failure diagnosis.

two kids playing on a teeter totter. One is named CO and the other is named O_2. As one kid goes up, the other goes down. Here are a few things which can cause an increase in CO which may help you understand the CO/O_2 relationship, and show how the balance between the two can be disrupted:

- A plugged air filter (A common cause of high CO on carbureted engines)
- A leaking fuel injector
- High fuel pressure
- A sticking carburetor choke
- A high carburetor float level
- A fuel saturated charcoal canister
- A faulty input to a vehicle computer which fools the PCM into thinking that a richer mixture is called for when in fact it is not.
- A contaminated crankcase that allows the PCV system to draw unburned fuel vapors into the intake system, richening the air-fuel mixture.

There are other possibilities, but the point is clear: a *decrease* in air or an *increase* in fuel will cause an increase in CO.

Low CO indicates a lean mixture

Correspondingly, anything which *increases* the flow of air, or *decreases* the flow of fuel to the combustion chamber will lean the mixture, causing a decrease in the amount of CO produced during combustion. CO has traditionally been measured as a percentage of total exhaust volume.

Hydrocarbons

Since unburned fuel *is* a hydrocarbon, this makes hydrocarbon emissions quite different from CO emissions, which only occur as a by-product of combustion. This is important, since an I/M 240 five-gas analyzer will detect the presence of hydrocarbons when placed near an open gasoline fuel filler neck or even near a windshield washer bottle filled with a washer fluid that contains a hydrocarbon-based additive, even though no combustion is taking place. Hydrocarbons are measured in ppm, or parts per million of sampled exhaust.

Hydrocarbon emissions are raw fuel. When sensed by an emission analyzer they are in vapor form, but they are raw fuel nonetheless. The inherent inefficiencies of the internal combustion engine will produce some HC. This is especially problematic when "cool" spots in the combustion chamber away from the spark plug quench the combustion process. When the flame goes out completely, we no longer get CO, but unburned HC still exits. The engine expels these hydrocarbons either through the exhaust, or by condensing them back into a liquid that leaks past the rings and into the crankcase.

Unburned fuel which condenses on the cylinder walls and runs back into the crankcase will eventually contaminate the crankcase to the point where the PCV system draws the raw fuel directly from the crankcase into the intake manifold. This will cause high CO and may result in high HC if the mixture gets too rich to burn, resulting in a misfire.

HC and incomplete combustion

High HC emissions also can be caused when any part of the combustion process fails, resulting in incomplete combustion.

Any item in our list of the components of good combustion can cause high HC emissions if they fail. An ignition failure such as a bad spark plug is one example. A mechanical failure such as a burned valve can lower compression and stop combustion, resulting in high HC emissions. At other times, a combined failure in the air-fuel delivery system, the engine mechanical, or the spark ignition system can cause high HC.

HC and ignition timing

Ignition timing affects HC emissions. Overly advanced ignition timing increases HC emissions by igniting the air-fuel mixture too early in the compression stroke. When the fuel charge is ignited before it is properly compressed, incomplete combustion occurs. Extremely retarded timing can also create high HC because it does not allow enough time for the fuel charge to burn before the exhaust valves open. Retarded timing stops combustion too early and forces the partially burned fuel charge, high in HC content, into the exhaust.

A relationship between CO and HC

High levels of either CO or HC in an exhaust sample are symptomatic of incomplete combustion. Extreme imbalances in CO (either very rich or very lean mixtures) can cause high HC emissions. This occurs when the mixture is too rich or too lean to support combustion.

At the rich side, if CO gets too high, HC will increase as combustion efficiency decreases or stops because of the overly rich mixture. A rich misfire will also cause an increase in O_2. You are probably familiar with wet spark plugs and misfires caused by a rich mixture.

The opposite happens when CO goes too low. This indicates a lean mixture which can no longer support good combustion. We refer to this as a "lean misfire". Once again, HC and O_2 increase as combustion efficiency decreases to the point of misfire.

HC and engine mechanical condition

Engine mechanical problems should always be eliminated as a potential cause for high HC before looking at more complicated or obscure causes for high HC. Remember that while the air-fuel mixture must be present in the correct amount, it must also be properly atomized and compressed for good combustion to occur. This means that in addition to air-fuel and good spark arriving at the correct instant, the cylinder must be properly sealed (rings, valves, and head gasket) during the compression stroke. Failure of any of these components results in poor combustion that leads to an increase in HC.

Oxygen

Oxygen is a very important part of combustion that reduces emission levels by improving the efficiency of the catalyst. Oxygen is also a good diagnostic tool for evaluating how rich or lean the air-fuel mixture is. Because this gas is such an effective mixture indicator, the O_2 sensor of an electronic control system uses information about oxygen levels in the exhaust as a way to signal whether the fuel mixture is rich or lean.

High levels of oxygen in the exhaust cause the O_2 sensor to generate a low voltage signal to the PCM. The PCM interprets a low voltage signal as an indication of a lean condition and a request for additional fuel in the mixture. Remember, lean and low.

Typically, 450 millivolts is the middle, or cross over point, of an O_2 sensor's operating range. A signal that is below 450 millivolts is considered low. The PCM sees this as a request for more fuel and responds by commanding a richer mixture. A lean mixture produces excess oxygen because there is not enough fuel available to chemically join with the oxygen during combustion. The result is left over oxygen in the exhaust after combustion is complete.

Low levels of oxygen in the exhaust cause the O_2 sensor to generate a high signal, above 450 millivolts, as an indication of a rich mixture. The PCM interprets a high signal as a rich condition and responds by reducing the amount of fuel in the mixture. With a high signal, there is too much fuel for the available oxygen in the intake charge. Most of the available oxygen in the cylinder is consumed during combustion, so there is very little left over in the exhaust.

However, be careful; the O_2 sensor can be fooled. It sees only oxygen as an indicator of a rich or lean condition. *The O_2 sensor does not directly measure CO!* The following are some potential sources of false oxygen, which can fool the sensor into "thinking lean":

- The oxygen content of any false air drawn into the exhaust ahead of the O_2 sensor will fool it into thinking that the mixture is lean. This air may be drawn directly into the engine through a vacuum leak, or enter the exhaust system ahead of the oxygen sensor through an exhaust leak, faulty check valve in the air injection system, or a cracked exhaust manifold.
- Cars equipped with some form of air injection can also pump false air into the exhaust ahead of the O_2 sensor. Some systems pump air into the exhaust manifold when the engine is cold to better oxidize the rich, unburned fuel in the exhaust. When the engine warms, this upstream air injection should stop. If it does not, it fools the O_2 sensor into thinking the mixture is too lean.
- A misfire will send unburned O_2 and hydrocarbons into the exhaust. The O_2 sensor seeing only the O_2, may think the mixture is lean, and richen the mixture.
- A misfire caused by a plugged fuel injector can shut off the fuel to a cylinder. Then the cylinder acts only as an air pump. The individual oxygen from this cylinder can fool the O_2 sensor into thinking that the entire engine is lean.
- Nominal readings of oxygen measured at the tailpipe should be 1.5 percent or less with air injection off, and 3 to 4 percent with secondary air turned on.

Carbon Dioxide

CO_2 is not used to pass or fail a car undergoing an emission test, but it is a very useful diagnostic tool. Since CO_2 is formed during complete combustion, it is an indicator of combustion efficiency. During good combustion, the carbon in the fuel combines with two molecules of oxygen to form the harmless CO_2 instead of the dangerous CO.

The higher the CO_2 level, the better. For engines without catalytic converters, a CO_2 level of 12 to 13 percent or more in the exhaust indicates good combustion efficiency. A working catalyst will raise CO_2 levels even higher. CO_2 levels slightly above 15 percent are not uncommon under these circumstances.

Oxides of Nitrogen

Our atmosphere is 78 percent nitrogen, 21 percent oxygen, and 1 percent other gases. This mix enters the combustion chamber as part of the air-fuel mixture. When combustion temperatures reach 2500°F, the nitrogen combines with oxygen to form oxides of nitrogen, or NO_x.

There are several contributing factors to the formation of NO_x, but all are related to high combustion temperatures. It is possible for cylinder temperatures to reach as high as 4500°F. Anything that raises combustion temperature to 2500°F or higher results in NO_x formation, figure 6-3. Common causes for these high temperatures group into three main categories, malfunctioning emissions equipment, incorrect tuning parameters, engine mechanical problems.

- Malfunctioning emissions equipment:
  1. Inadequate EGR function, allowing combustion chamber temperatures to rise above 2500°F.
  2. A malfunctioning preheat door on the air cleaner intake. When the intake flapper door sticks in the hot position, only preheated manifold air can enter the engine. When the engine is cold, this does not cause a problem, but when the engine is hot, the preheated air can raise heat from compression to a point where combustion temperatures go above the 2500 degree mark.
  3. The Early Fuel Evaporation (EFE) system adding heat to the air inside the intake manifold.
  4. A defective 3-way catalytic converter.
  5. Too much O_2 injected upstream of the catalytic converter by the air injection system (AIR). The reduction reaction in the converter removes oxygen from the NO_x to produce harmless nitrogen and carbon dioxide. This works most efficiently when some CO is present but there is not an excess of O_2.
- Incorrect tuning parameters:
  1. Advanced timing caused by setting it incorrectly or by a faulty mechanical or vacuum advance unit. The computer may also advance the timing too far if a:
     a. Faulty coolant temperature sensor indicates a cold engine.
     b. Faulty BARO sensor indicates a higher than correct altitude.
     c. Knock Sensor system is faulty.
     d. EGR position sensor indicates excess EGR.
     e. PCM or PROM is faulty.
  2. Very retarded timing. Slightly retarded timing reduces engine temperatures and NO_x. However, very retarded timing reduces engine efficiency, causes the engine to work harder, raising combustion temperatures and NO_x emissions.
  3. A lean fuel mixture. Lean fuel mixtures raise combustion temperatures. They also lead to an excess of O_2 in the exhaust, which reduces the effectiveness of the reduction reaction in the catalytic converter.
- Engine mechanical problems:
  1. High engine temperatures caused by a stuck thermostat or plugged cooling system.
  2. A partially plugged cooling system. Obstructions to coolant flow caused by deposits right next to the combustion chamber can cause isolated hot spots even though the average coolant temperature may not register a problem in the dashboard temperature gauge.
  3. Carbon deposits inside the cylinder reduce the size of the combustion chamber, raising compression. An increase in compression causes an increase in temperature.

BUILDING A STOICHIOMETRIC CHART

We have viewed each of the five gases separately. Along the way we have shown a few relationships between several of the gases. Now it is time to put them all together in one chart. We will start back at the beginning and build the chart slowly. When we are done, we will have a stoichiometric chart.

Stoichiometry

Even if you have been exposed to the concept of stoichiometry before, and understand the concept, we need to impress upon you just how important it is to the proper operation of a modern fuel system and its components. Understanding stoichiometry and the relationships between the gases is essential to using exhaust gas readings as a diagnostic tool. When we are done, a look at our charts will convince you that the stoichiometric air-fuel ratio of 14.7:1 was not arrived at by chance.

CO and O_2 — rich-lean indicators

Remember, these two gases are our rich-lean indicators, figure 6-4. On the rich side, to the left of our chart, CO is high, but O_2 is low. We have a lot of fuel and not enough air over on the rich end. As we near stoichiometry, however, we see the two gases coming together.

O_2 started low on the rich side of the chart, and CO was high, since there is plenty of fuel available to combine with the oxygen when a rich mixture is present. But watch what happens as the mixture starts to lean out as it approaches stoichiometry. Right in the middle at 14.7 to 1 we can see that CO and O_2 come into balance and the lines representing each intersect. In a balanced fuel mixture, CO and O_2 are evenly balanced.

As soon as we get past 14.7 to 1 and start moving toward the lean side of the chart, we see that CO and O_2 begin to swap the roles they played on the rich side of the chart. They start moving away from one another again, but this time in the opposite directions.

	EXHAUST EMISSIONS				
ENGINE CONDITIONS	HC	CO	CO_2	O_2	NO_x
Ignition misfire	Large Increase	Some Decrease	Some Decrease	Some-Large Increase	Some-Large Decrease
Compression loss	Some-Large Increase	Some Decrease	Some Decrease	Some Increase	Some-Large Decrease
Rich mixture	Some-Lar Increase	Large Increase	Some Decrease	Some Decrease	Some-Large Decrease
Lean mixture	Some Increase	Large Decrease	Some Decrease	Some Increase	Some-Large Decrease
Very lean mixture	Large Increase	Large Decrease	Some Decrease	Large Increase	Some Increase
Slightly retarded timing	Some Decrease	No Change or Some Increase	No Change	No Change	Large Decrease
Very retarded timing	Some Increase	No Change	Some-Large Decrease	No Change	Some Increase
Advanced timing	Some Increase	No Change or Some Decrease	No Change	No Change	Large Increase
EGR operating	No Change	No Change	Some Decrease	No Change	Large Decrease
EGR leaking	Some Increase	No Change	No Change or Some Decrease	No Change	Some Decrease or No Change
Air Injection operating	Large Decrease	Large Decrease	Some-Large Decrease	Large Increase	No Change
Catalytic converter functional	Some Decrease	Some Decrease	Some Increase	Some Decrease	Some Decrease w/3-w cat
Exhaust leak	Some Decrease	Some Decrease	Some Decrease	Some Increase	No Change
Worn Engine	Some Increase	Some Increase	Some Decrease	Some Decrease	No Change or Slight Decrease
Flat Camshaft	No Change or Some Decrease	Some Decrease	Some Decrease	No Change or Some Decrease	No Change or Some Decrease
	ENGINE CONDITION'S EFFECT ON EXHAUST EMISSIONS				

Figure 6-3. These 5-gas readings can help with emission failure diagnosis.

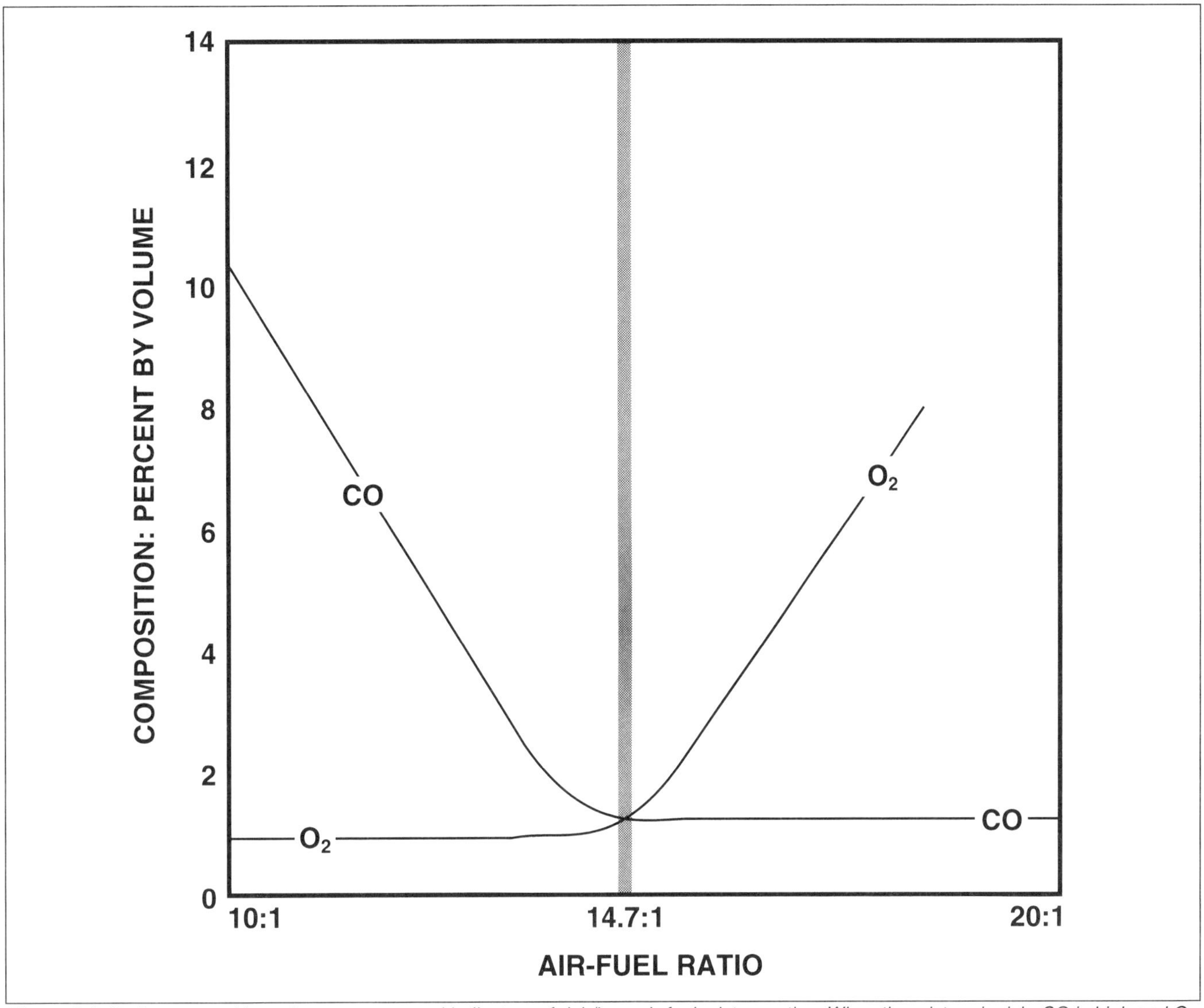

Figure 6-4. Carbon monoxide and oxygen are good indicators of rich/lean air-fuel mixture ratios. When the mixture is rich, CO is high and O_2 is low. When the mixture is lean, CO is low and O_2 is high.

On the lean side, CO stays low, too low to measure with the exhaust gas analyzer. But without any fuel to burn, we end up with a lot of left over oxygen which alerts us to the lean condition. The nice part about oxygen, is that it increases very rapidly as the mixture leans past stoichiometry. This, plus the fact that it is less affected by the catalytic converter, makes oxygen much better than CO as an indicator of a lean condition.

Finally, remember that while there is always some oxygen, even with a rich condition, there cannot be any CO in the exhaust without combustion. Since catalyzed levels of CO at the tailpipe on a car with a catalyst may not even register on a gas analyzer, O_2 becomes the best diagnostic tool for identifying a lean mixture.

Hydrocarbons — raw fuel

Anytime we have incomplete combustion, we have leftover, unburned fuel or HC, figure 6-5. The air-fuel mixture has a direct affect on the amount of hydrocarbons in the exhaust, but it is not the only thing that affects HC levels. When the air-fuel ratio varies too much from the stoichiometric ideal, the mixture becomes either too rich or too lean to burn properly. Both situations result in high levels of HC in the exhaust.

An extremely rich mixture will cause an excessively high HC concentration. In this type of super-rich condition, spark plugs foul and black smoke is visible at the tailpipe. This extreme condition does not happen instantaneously, HC will increase gradually between stoichiometry and the fouled plug stage.

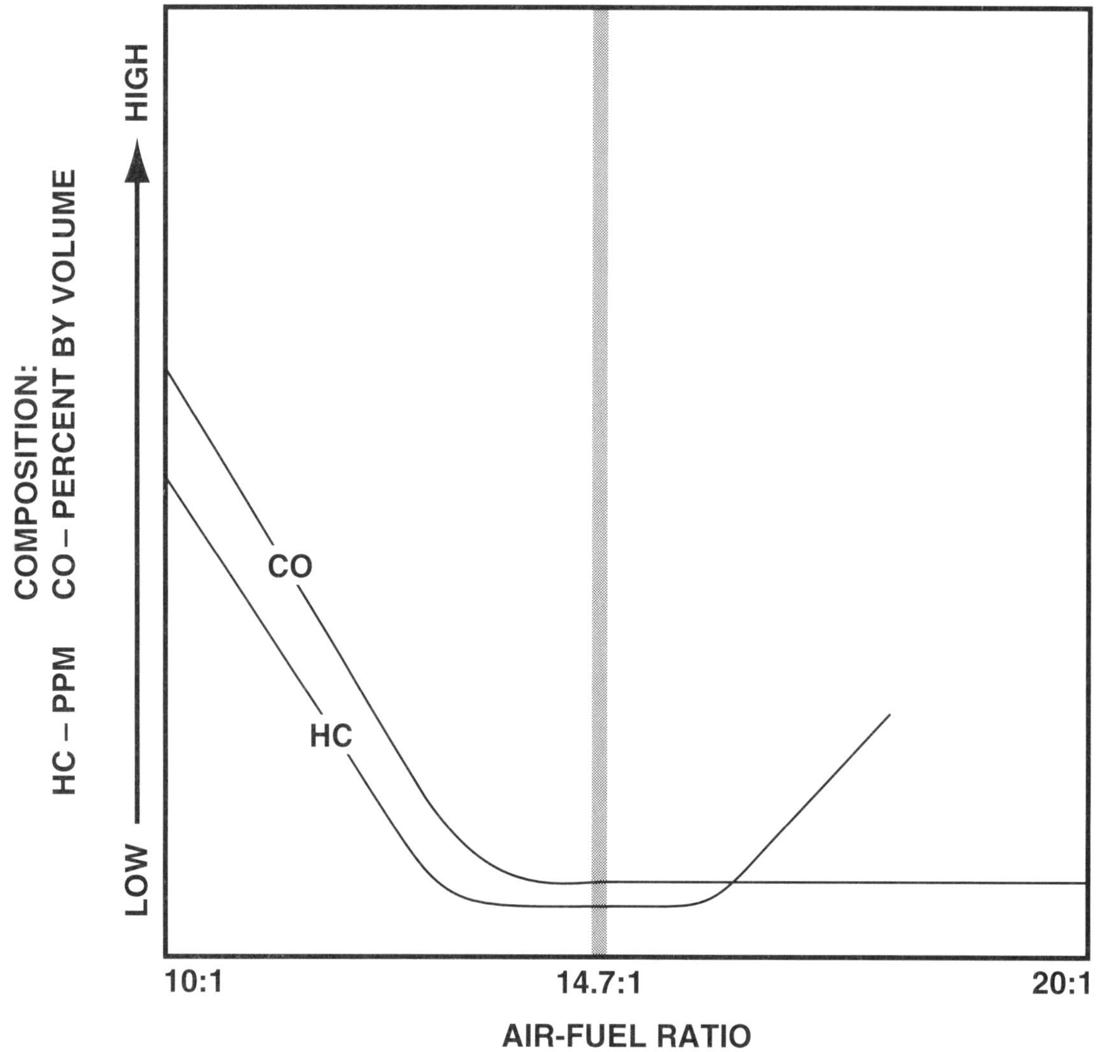

Figure 6-5. The hydrocarbon emissions go up if the air-fuel ratio becomes either too rich or too lean.

A very lean mixture will also cause high HC concentrations. As the mixture leans out, it will get to a point where there just is not enough fuel to support combustion. We call this condition a *lean miss.*

CO_2 — indicator of combustion efficiency

So far, we looked at HC, CO, and O_2 as indications that our engine is running at stoichiometric efficiency. When any one of these three is present in the exhaust in higher than normal levels, it is an indication of *poor* combustion efficiency. Ideally, levels of all three should be low if the engine is running properly.

CO_2 is the one gas that should be *high*, the higher the better, figure 6-6. Unlike the three bad gases, (HC, CO and NO_x), CO_2 is not a pollutant. Since it is only formed during complete combustion, CO_2 is a great indicator of overall combustion efficiency. A high level of CO_2 tells us that all of the elements of complete combustion are present, from good compression to a balanced air-fuel ratio, to strong ignition. Notice that when CO_2 is highest near stoichiometry, CO and O_2 are both low and evenly balanced.

Some sources suggest that 8 percent is a minimum allowable CO_2 level. It is probably more useful and realistic to remember that a good running engine, operating at stoichiometry should produce over 13 percent CO_2, and that CO_2 levels will actually increase above this point as a result of catalyst action. This makes CO_2 an indicator of both engine efficiency and catalyst action.

Stoichiometry and NO_x

A stoichiometric air-fuel mixture is also beneficial for the reduction of NO_x emissions. At air-fuel ratios leaner than 14.7:1, NOx emissions begin to rise dramatically. Rich mixtures, less than 14.7:1, burn cooler so

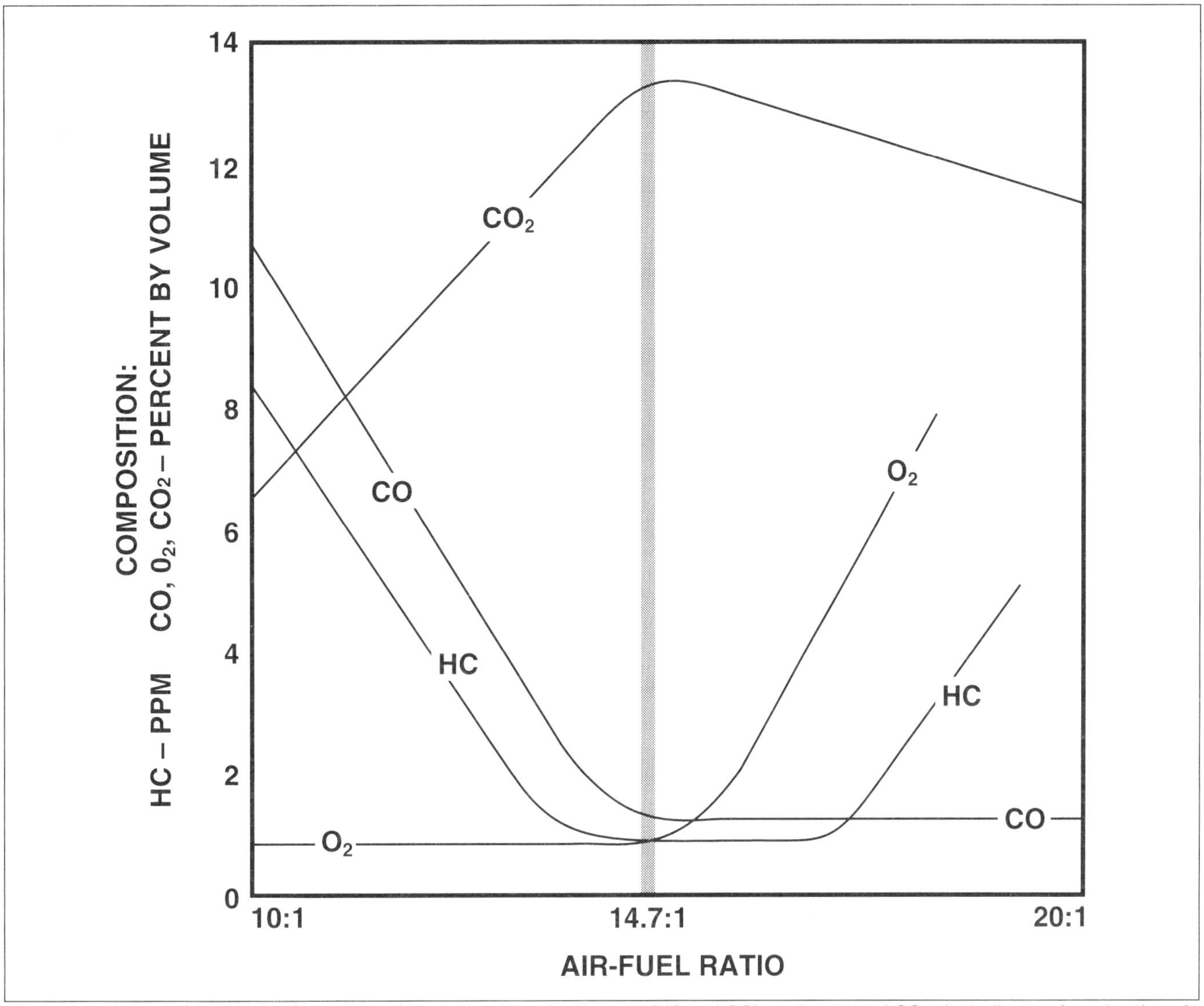

Figure 6-6. At a 14.7:1 air-fuel ratio, emission is optimum. The "bad" gases (HC and CO) are lowest and CO_2, the indicator of combustion efficiency is at its highest level.

combustion temperatures stay below 2500°F and NO_x emissions remain low as well. A very rich air-fuel ratio is an effective NO_x reducer, but such a ratio results in higher HC and CO emissions, as well as reduced fuel economy. Obviously, this is an unaccpectable trade, so 14.7:1 is the best compromise. The exhaust gas recirculation (EGR) system eliminates most of the NO_x emissions that can form in the combustion chamber at the stoichiometric ratio.

The reduction reaction in a three-way catalytic converter further reduces NO_x emissions, figure 6-7. Fortunately, converters work best when the air-fuel ratio is 14.7:1, figure 6-8. Too rich a mixture will overwork a catalytic converter, and extended rich running will cause converter overheating and failure.

Although new vehicles have been required to meet a federal NO_x emissions standard for quite some time, monitoring NO_x during in service vehicle testing has not been required until recently. To pass an I/M 240 emissions test, a vehicle must meet tailpipe standards for a fifth gas, NO_x. Remember, NO_x can only form when combustion temperatures are high, and either a lean mixture or a lack of EGR are usually responsible. Since the EGR system only operates at cruise, it cannot cause high NO_x emissions at idle. Levels of the other four gasses will verify a lean condition.

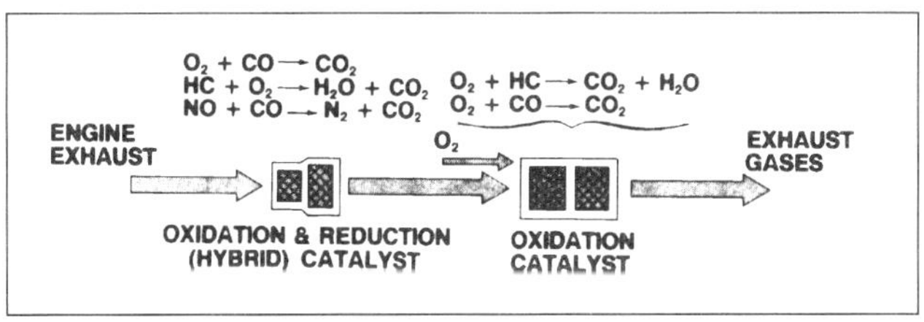

Figure 6-7. A catalytic converter reduces emissions by oxidation and reduction reactions.

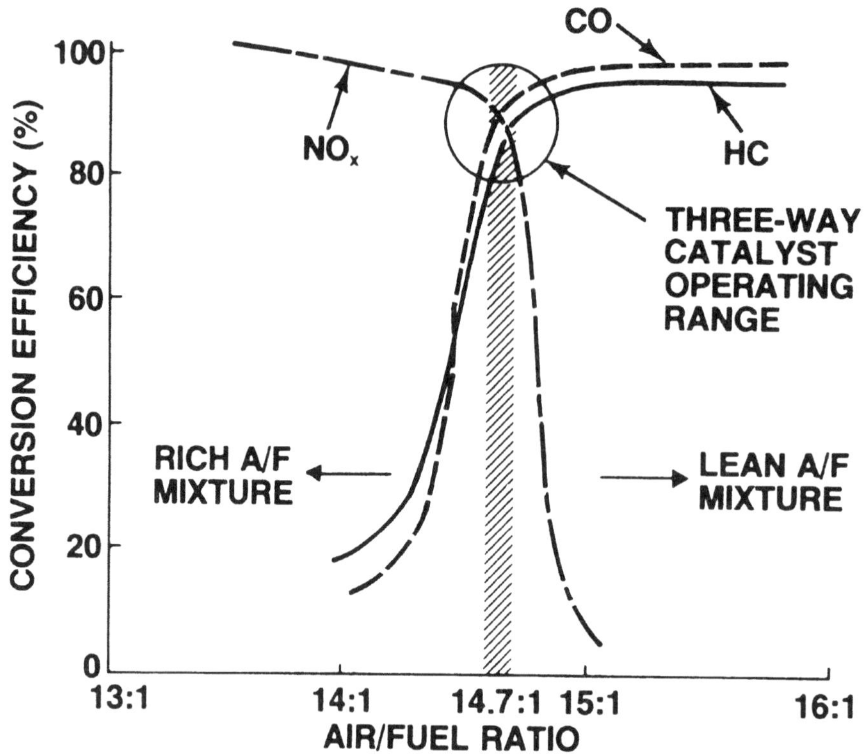

Figure 6-8. A 3-way catalytic converter most efficiently reduces emissions when the air-fuel ratio is held with the narrow range of 14.65:1 and 14.75:1.

1. Which should be true of a vehicle's stoichiometric air-fuel ratio?
   a. HC is low; CO, O_2, and CO_2 are high
   b. HC and CO are low; O_2 and CO_2 are high
   c. HC, CO, and O_2 are low; CO_2 is high
   d. HC, CO, O_2, and CO_2 are low

   Explanation: HC and CO are both deadly gases and they should be low at ideal combustion. O_2 should also be fairly low. If it is too high, there will be too much air and a lean misfire will occur. If there is very little or no O_2 the mixture is rich and emissions will increase. CO_2 should always be high for stoichiometry to occur. In general, the higher the CO_2 the better. With perfect combustion the byproduct is CO_2 and water. Therefore, c is the best answer.

2. The composite vehicle fails the I/M 240 emissions test due to excessive NO_x levels.

   Technician A says this could be caused by a broken vacuum hose to the EGR valve.

   Technician B says this could be caused by an open at pin #47 to the PCM.

   Who is right?
   a. A only
   b. B only
   c. Both A and B
   d. Neither A nor B

   Explanation: NO_x emissions will increase anytime the EGR is not able to turn on. In both problems mentioned by the technicians above, the EGR valve would be prevented from opening. The PCM uses pin #47 to modulate the EGR valve on and off. With an open circuit, it would be incapable of energizing the EGR solenoid. It would also prevent the EGR from turning on whenever there is a loss of vacuum, so the best answer is c.

3. During an emission test, the composite vehicle fails the test due to high HC readings at idle. This can be caused by all of the following EXCEPT:
   a. An open spark plug wire
   b. A fouled spark plug
   c. Carbon build up in the IAC motor valve assembly and throttle body
   d. A fault in the VSS circuit

   Explanation: We are looking for what will not cause a vehicle to fail the emissions test due to a high HC reading at idle. The VSS circuit does not have any influence on the emissions while idling in park. An open or fouled spark plug will clearly cause a misfire and high HC levels. A build up of carbon in the idle air control motor and/or throttle body may cause higher HC due to low and erratic idle speed. The best answer is d because the VSS will not have any effect on the vehicle's emissions at idle.

4. High HC exhaust emissions are often due to:
   a. An overheated engine
   b. A restricted air cleaner
   c. Ignition system problems
   d. An inoperative EGR system

   Explanation: Ignition system, or c, is the best answer. Any time the ignition system fails to fire the spark plug for any number of reasons, the result is high H/C (unburned fuel).

5. The composite vehicle fails the idle portion of the I/M 240 test. The engine idles roughly and runs smoothly as the engine speed increases. With the information shown below, what is the most likely cause?
   a. An EGR valve that is stuck open
   b. A shorted injector driver
   c. A vacuum leak at the intake manifold
   d. A fouled spark plug

   Explanation: A leaking EGR valve forces inert exhaust gas to displace the fresh air and fuel entering the cylinders, so there is less air and fuel to burn. This results in very poor combustion and incomplete burning of fuel which is most critical at idle. Therefore, the HC level is high due to the unburned fuel and the CO level is low because of the dilution of the burnable air and fuel mixture entering the combustion chamber. The CO_2 percent is low due to the poor combustion efficiency. The percent of oxygen remained low at idle since the EGR gas displaced a portion of it. All of the other conditions listed would have caused higher than normal oxygen readings during idle, so a is the best answer.

ENGINE SPEED	IDLE	2,000 RPM
HC (PPM)	600	25
CO (PERCENT)	0.40	0.20
CO_2 (PERCENT)	12.0	14.5
O_2 (PERCENT)	0.50	0.90

Illustration for question 5.

6. The probe of an exhaust analyzer has been placed near the open neck of a radiator on a running engine suspected of having a blown head gasket.

   Technician A says that a high HC reading will indicate a leak between the combustion chamber and the cooling system.

   Technician B says that a high CO reading is a better indicator of a blown head gasket.

   Who is right?

   a. A only

   b. B only

   c. Both A and B

   d. Neither A nor B

   Explanation: Technician A could be sampling HC from a number of sources, including the windshield washer bottle, a leaking canister purge hose, or any other source of raw HC present under the hood. Technician B is correct. Since the engine is running, and since CO is formed only during combustion, its presence in the cooling system would be a more accurate indicator of a leak between the combustion chamber and cooling system.

7. A vehicle which passes for both HC and CO, fails the I/M 240 emission test for NO_x. What is the cause?

   a. A plugged EGR port

   b. A carbon buildup in the piston heads

   c. Hot spots in the combustion chamber caused by a cooling system blockage in the water jacket surrounding the cylinders.

   d. All of the above

   Explanation: Anything which increases the combustion temperature to 2500°F or more can cause NO_x formation. Since a plugged EGR port would stop the flow of exhaust gas to the cylinder, it wouldn't introduce any exhaust gas to the cylinder to cool the combustion process. Carbon buildup on the piston heads would reduce the size of the combustion chamber, increasing compression. Since higher compression produces higher combustion temperatures, the carbon could be raising the combustion chamber temperature above 2500°F. The same would apply to hot spots in the water jacket. Localized hot spots in the cooling system, especially right at the water jacket surrounding the cylinder, could cause an increase of combustion temperature. The answer is d.

8. An engine idles well, but it is emitting higher than acceptable levels of HC. CO is within specs.

   Technician A says that over-advanced ignition timing could be causing the high HC.

   Technician B says a plugged air filter could be causing the high HC.

   Who is right?

   a. A only

   b. B only

   c. Both A and B

   d. Neither A nor B

   Explanation: Technician B has forgotten that the CO level is correct. A plugged air filter would have caused an increase in CO since the restricted flow of air would drive the mixture rich. Technician A realizes that over-advanced timing may not affect the idle quality of the engine unless it is extremely over advanced. But a few degrees of timing advance can cause a substantial increase in HC emissions. Since the spark is occurring too soon in the combustion stroke, most of the heat of combustion is concentrated right at the piston head. This lowers the temperature of the cylinder walls, and quenching of the flame front results in less complete combustion. Additionally, the high concentration of heat at the piston, means there is less heat exiting with the exhaust. Under ideal conditions, high heat in the exhaust will continue the burning process in the exhaust, burning HC which was left over from the combustion process.

9. A car fails the emission test for both high HC and CO. Oxygen readings at the tailpipe are above 5 percent. An examination of the vehicle shows that the catalyst is empty. With a new catalyst installed, it is discovered that the oxygen sensor voltage never goes above 200 millivolts.

   Technician A says that the oxygen sensor is bad, and should be replaced.

   Technician B says that the O2 sensor may be good, but a leak at the exhaust manifold is fooling the sensor into sending a lean signal.

   Who is right?

   a. A only

   b. B only

   c. Both A and B

   d. Neither A or B

   Explanation: Both technicians *could* be correct in this instance. Sure its possible that the oxygen sensor has failed for the same reason the catalyst burned up. But Technician A does not know that for sure. Technician B is a lot closer to making the correct fix, because he is using the information available to him to list possible causes for the failure. He knows the O2 sensor is signalling a lean mixture, but he also knows that there is a lot of oxygen exiting the tailpipe, oxygen which is entering the exhaust *ahead* of the O2 sensor at the leaking exhaust manifold gasket. A new O2 sensor will not eliminate the false oxygen, so a new sensor will still signal a lean mixture, which will drive the system rich and destroy the new catalyst. His approach is better; fix the exhaust leak first, and retest the oxygen sensor before condemning it.

10. A vehicle emits high HC during an emission test. Catalyzed CO is too low to measure, and O2 sensor voltage is below 250 millivolts. Propane is used to richen the mixture through the air inlet snorkel at the air cleaner housing. This causes CO to rise to measurable levels and HC drops. O2 sensor voltage increases to 850 millivolts.

    What is the most likely cause of the original high HC readings?

    a. a leaking brake booster diaphragm

    b. a fouled spark plug

    c. a high float level

    d. a dirty air filter element

    Explanation: Since the O2 sensor voltage and CO both rise together with propane enrichment, we know that the O2 sensor is working, and that its original low voltage signal properly indicated a lean mixture. The corresponding decrease in HC tells us that the high hydrocarbon readings were the result of a lean miss—the type of miss that would be caused by a vacuum leak. A ruptured brake booster diaphragm would cause such a leak, the answer is a. A fouled spark plug is certainly one possible cause for high HC, but enriching the mixture wouldn't have corrected that problem. The plug would still be fouled, and HC wouldn't drop with a richer mixture. A high float level or plugged air filter would both contribute to a rich mixture and high CO, a condition which would only worsen with propane enrichment.

SCAN TOOL DATA			
Engine Coolant Temperature Sensor **(ECT)** 0.45 Volts	Intake Air Temperature Sensor **(IAT)** 2.40 Volts	Manifold Absolute Pressure Sensor **(MAP)** 0.60 Volts	Throttle Position Sensor **(TPS)** 0.50 Volts
Engine Speed Sensor **(RPM)** 850 RPM	Heated O2 Sensor **(HO2S)** 0.05 Volts	Vehicle Speed Sensor **(VSS)** 0.0 MPH	Battery Voltage **(B+)** 13.9 Volts
Idle Air Control Valve **(IAC)** 20 Percent	Evaporative Emiss Canister Purge Sol **(EVAP)** On **Off**	Torque Converter Clutch Solenoid **(TCC)** On **Off**	EGR Valve Control Solenoid **(EGR)** 0.0 Percent
Malfunction Indicator Lamp **(MIL)** **On** Off	Diagnostic Trouble Codes **(DTC)** P0171	Open/Closed Loop **Open** Closed	Fuel Pump Relay **(FP)** **On** Off
Measured Ignition Timing °BTDC	Base Timing: 10	Actual Timing: 8	

Illustration for question 1.

1. The composite vehicle comes into your shop because it has failed an I/M test due to a rich condition. It has the scan tool readings shown above. The injectors have a wider than normal pulse width and O2 sensor voltage stays low. This could be caused by:

   a. A fault in the ECT sensor circuit.

   b. A bad O2 sensor or its circuit.

   c. A fault in IAT sensor circuit.

   d. A faulty TP sensor.

   Explanation: All scan tool data is within limits except the O2 sensor. The O2 sensor voltage is staying low (lean) when it should be higher because of a rich condition. When the O2 sensor is malfunctioning and producing very little voltage, the PCM thinks the engine is running lean so it sends a rich command to the injectors. The PCM is not aware that the oxygen sensor is still not able to produce enough voltage for a rich command, so in turn, it adds more fuel and sets a trouble code PO171 system (system too lean). The correct answer is b. (In some cases it will also set a code P0130 (O2S circuit malfunction)).

2. A fully OBD II compliant vehicle is being tested. The customer is complaining of a check engine light coming on. The trouble code P0440 was stored in the PCM's memory.

   a. This is caused by an EGR valve malfunction.

   b. An open or short in the canister purge solenoid could be the cause.

   c. This could be caused by a plugged or leaking canister purge hose.

   d. Both b and c

   Explanation: The trouble code P0440 will be set on a fully OBD II vehicle when it detects a fault in the canister purge system regardless of whether it is electronically or mechanically caused. Therefore, both b and c are correct, so the answer is d.

3. Technician A says that most MAP and TPS sensors operate on a 5.0 volt reference.

   Technician B says that most MAP and TPS sensors are analog signals which vary from approximately 0.40 to 1.1 volts at idle to a maximum of almost 5.0 volts at full throttle.

   Who is right ?

   a. A only

   b. B only

   c. Both A and B

   d. Neither A nor B.

   Explanation: Most vehicles, including the composite vehicle, use a 5.0 volt reference to feed the MAP and the TPS sensors. Both the TPS and MAP sensors will operate at approximately 0.40 to 1.10 volts at idle and close to the reference voltage of 5.0 volts at full throttle. Since both Technicians are right, the best answer is c.

4. The composite vehicle is being test driven on the highway while passing another car, the engine begins to cut out at approximately 6,000 rpm's.

   Technician A says a bad ignition module is most likely the cause.

   Technician B says a restricted fuel filter could be the cause.

   Who is right?

   a. A only

   b. B only

   c. Both A and B

   d. Neither A nor B

   Explanation: You must be careful answering this question, since all of the answers could be true. The reason d is the best answer is because in the composite vehicle reference booklet supplied by ASE, it states "The PCM will turn off the fuel injectors, for safety reasons, when the vehicle speed reaches 110 mph, or if the engine speed exceeds 6,000 rpm's". It is important to understand this information and carefully read the composite vehicle reference booklet.

SCAN TOOL DATA			
Engine Coolant Temperature Sensor **(ECT)** 0.60 Volts	Intake Air Temperature Sensor **(IAT)** 2.50 Volts	Manifold Absolute Pressure Sensor **(MAP)** 0.0 Volts	Throttle Position Sensor **(TPS)** 0.0 Volts
Engine Speed Sensor **(RPM)** 1,500 RPM	Heated O2 Sensor **(HO2S)** 0.85 Volts	Vehicle Speed Sensor **(VSS)** 20 MPH	Battery Voltage **(B+)** 13.5 Volts
Idle Air Control Valve **(IAC)** 20 Percent	Evaporative Emiss Canister Purge Sol **(EVAP)** On Off	Torque Converter Clutch Solenoid **(TCC)** On Off	EGR Valve Control Solenoid **(EGR)** 0.0 Percent
Malfunction Indicator Lamp **(MIL)** On Off	Diagnostic Trouble Codes **(DTC)** P0105, P0120	Open/Closed Loop Open Closed	Fuel Pump Relay **(FP)** On Off
Measured Ignition Timing °BTDC	Base Timing: 10	Actual Timing: 15	

Illustration for question 5.

5. The composite vehicle intermittently runs rough, stalls and hesitates, then turns the MIL light on. The scan tool data shown above was captured during the duplication of this problem. This condition could be caused by a loose, poor or damaged connection at:
   a. PCM pin #16
   b. PCM pin #13
   c. PCM pin #22
   d. PCM pin #21

   Explanation: When pin #13 goes open, there is not any voltage to the TPS or MAP sensors. The reason the ECT and the IAT sensor voltages were not affected is because they receive their own 5.0 volt reference directly from the PCM and only use the signal return line from pin #16. The best answer is b.

6. The composite vehicle has been towed to the shop. The customer complains that the engine starts normally but dies as soon as it is shifted into drive or reverse. Using the scan tool data shown below, as well as the composite vehicle information, determine which of these is the most likely cause?
   a. A plugged catalytic converter
   b. Pin #48 is shorted to ground
   c. An open circuit at pin #45
   d. A high resistance at pin #47

SCAN TOOL DATA			
Engine Coolant Temperature Sensor **(ECT)** 0.60 Volts	Intake Air Temperature Sensor **(IAT)** 2.50 Volts	Manifold Absolute Pressure Sensor **(MAP)** 0.80 Volts	Throttle Position Sensor **(TPS)** 0.55 Volts
Engine Speed Sensor **(RPM)** 850 RPM	Heated O2 Sensor **(HO2S)** 0.50 Volts	Vehicle Speed Sensor **(VSS)** 0.0 MPH	Battery Voltage **(B+)** 13.8 Volts
Idle Air Control Valve **(IAC)** 50 Percent	Evaporative Emiss Canister Purge Sol **(EVAP)** On Off	Torque Converter Clutch Solenoid **(TCC)** On Off	EGR Valve Control Solenoid **(EGR)** 0.0 Percent
Malfunction Indicator Lamp **(MIL)** On Off	Diagnostic Trouble Codes **(DTC)** P0740	Open/Closed Loop Open Closed	Fuel Pump Relay **(FP)** On Off
Measured Ignition Timing °BTDC	Base Timing: 10	Actual Timing: 12	

Illustration for question 6.

Explanation: With pin #48 TCC shorted to ground, the torque converter clutch will energize and the clutch will be applied regardless of vehicle speed or coolant temperature. As soon as the transmission is shifted into drive or reverse with the brake applied, the engine will die. Since trouble code PO740 was stored in the computer memory, it should have clued you to look carefully at the TCC circuit. The answer is b.

7. The composite vehicle has failed the I/M 240 emissions test. The customer complains that ever since **YOU** did the last brake job last year, his car has begun to run worse and worse. It is very sluggish, gets poor fuel mileage, seems to loose power and now it not only stinks but the MIL light stays on. Use the information shown below, recorded with the engine at operating temperature, to diagnose the mostly likely origin of your customers problem.
   a. MAP sensor circuit
   b. TPS and IAT sensor circuits
   c. ABS hydraulic modulator
   d. ECT sensor or its circuit

   Explanation: The ECT circuit at 3.8 volts would make the PCM think the engine is very cold. This in turn would cause the car to run very rich. In time, this will damage the CAT, restricting the exhaust flow through the catalytic converter and cause a faulty MAP sensor reading. Thus, setting a MAP sensor code and a rich running code. The ECT sensor reading of 3.8 volts was not far enough out of limits to set a PCM code. The best answer is d.

EXHAUST EMISSIONS		
ENGINE SPEED	**IDLE**	**2,000 RPM**
HC (PPM)	**300**	**250**
CO (PERCENT)	**4.0**	**4.75**
CO_2 (PERCENT)	**10.8**	**10.5**
O_2 (PERCENT)	**0.50**	**0.45**

SCAN TOOL DATA			
Engine Coolant Temperature Sensor **(ECT)** 3.8 Volts	Intake Air Temperature Sensor **(IAT)** 2.5 Volts	Manifold Absolute Pressure Sensor **(MAP)** 1.5 Volts	Throttle Position Sensor **(TPS)** 0.50 Volts
Engine Speed Sensor **(RPM)** 1,050 RPM	Heated O2 Sensor **(HO2S)** 0.95 Volts	Vehicle Speed Sensor **(VSS)** 0.0 MPH	Battery Voltage **(B+)** 13.9 Volts
Idle Air Control Valve **(IAC)** 35 Percent	Evaporative Emiss Canister Purge Sol **(EVAP)** On Off	Torque Converter Clutch Solenoid **(TCC)** On Off	EGR Valve Control Solenoid **(EGR)** 0.0 Percent
Malfunction Indicator Lamp **(MIL)** On Off	Diagnostic Trouble Codes **(DTC)** P0172, P0105	Open/Closed Loop Open Closed	Fuel Pump Relay **(FP)** On Off
Measured Ignition Timing °BTDC	Base Timing: 10	Actual Timing: 10	

Illustration for question 7.

ENGINE SPEED	IDLE	2,000 RPM
HC (PPM)	30	20
CO (PERCENT)	0.35	0.15
CO_2 (PERCENT)	14.9	15.5
O_2 (PERCENT)	1.0	1.5

Illustration for question 8.

8. An I/M 240 emission test has just been performed on the composite vehicle (see reading above).

   Technician A says the composite vehicle obviously has a faulty TPS sensor.

   Technician b says that there is a problem with the O2 sensor or its circuit.

   Who is right?

   a. A only
   b. B only
   c. Both A and B
   d. Neither A nor B

   Explanation: There is nothing wrong here. Both Technicians are wrong, so d is the best answer. Anytime you have CO_2 readings as high as these, you have an efficiently running car. The rest of the readings look good as well.

9. After replacing the timing belt on a customers car, he comes back complaining that ever since you replaced it, the engine has run rough, surges and misfires. You tell your customer that you believe the belt was correctly installed, but you will check out the car and let him know what the problem is that you find.

   After checking the car on the scope, you see the following ignition secondary abnormality shown below.

   a. #2 spark plug wire has a high resistance.
   b. #2 spark plug is fouled.
   c. The coil is defective.
   d. The ignition module is faulty.

	DIGITAL SECONDARY KV			
CYL #	BURN TIME MS	AVERAGE KV	SNAP KV	BURN KV
1	1.5	9	14	1.5
2	0.30	38	41	2.9
3	1.7	8	13	1.4
4	1.6	10	17	1.5

Illustration for question 9.

Explanation: The high firing voltage of 38kv, and a short burn time as well as a high burn KV, is due to a high resistance in the secondary circuit of cylinder #2. So the best choice would be an open #2 spark plug wire, so the answer is a. If the spark plug was fouled, the voltages would have all been low with little or no change. The burn time MS would have been much longer on the affected cylinder. A bad coil would affect all cylinders and a faulty ignition module would clearly be a wrong answer.

10. The composite vehicle is towed in to your shop and will not start. The driver of the tow truck claims that it appears to have a fuel starvation problem. Using the scan tool data shown below, determine the most likely cause.

    a. Fill the gas tank!
    b. The faulty throttle position sensor.
    c. An open circuit at pin #16.
    d. A short at pin #13.

    Explanation: With an open circuit at pin #16 the MAP, TPS, ECT, and IAT sensors would lose their signal return. This would cause all the sensors to read close to the 5.0 volt reference. This alone would make the vehicle hard to start along with the fact that the PCM virtually shuts off the fuel injectors when it sees a high TPS voltage and engine RPM below 400. The answer is c.

SCAN TOOL DATA			
Engine Coolant Temperature Sensor **(ECT)** 4.91 Volts	Intake Air Temperature Sensor **(IAT)** 4.92 Volts	Manifold Absolute Pressure Sensor **(MAP)** 4.90 Volts	Throttle Position Sensor **(TPS)** 5.0 Volts
Engine Speed Sensor **(RPM)** 250 RPM	Heated O2 Sensor **(HO2S)** 0.0 Volts	Vehicle Speed Sensor **(VSS)** 0.0 MPH	Battery Voltage **(B+)** 10.2 Volts
Idle Air Control Valve **(IAC)** 20 Percent	Evaporative Emiss Canister Purge Sol **(EVAP)** On Off	Torque Converter Clutch Solenoid **(TCC)** On Off	EGR Valve Control Solenoid **(EGR)** 0.0 Percent
Malfunction Indicator Lamp **(MIL)** On Off	Diagnostic Trouble Codes **(DTC)** P0105, P0110, P0115 P0120, P0171, P0560	Open/Closed Loop Open Closed	Fuel Pump Relay **(FP)** On Off
Measured Ignition Timing °BTDC	Base Timing: 10	Actual Timing:	

Illustration for question 10.

11. Multiple code problems are often caused by a poor ground connection and the best way to check the ground circuit is to perform a voltage drop test. Voltage drop on the main PCM ground circuit should be:

    a. 0.01 volt or less

    b. 0.10 volt or less

    c. 0.50 volt or less

    d. 1.00 volt or less

    Explanation: Voltage drop on a ground circuit should be less than 0.10 volt. Therefore, b is the correct answer. A low resistance path to ground is critical, especially on computer-controlled circuits that operate on a low-voltage signal.

12. A fuel pressure check is being performed on the composite vehicle. The measured pressure at idle is 48.0 psi.

    Technician A says this could be caused by a restriction in the fuel return line.

    Technician B says this could be caused by a plugged vacuum hose to the fuel pressure regulator.

    Who is right?

    a. A only

    b. B only

    c. Both A and B

    d. Neither A nor B

    Explanation: Normal pressure for the composite vehicle is 34-38 psi. Restricting the fuel return line will cause the fuel pressure to increase. A lack of vacuum to the fuel pressure regulator due to a plugged hose will also cause pressure to increase, so both technicians are correct.

13. With the key on but the engine off, a technician presses the throttle to over 90 percent of its opening. What voltage would a voltmeter display at pin #23 of the composite vehicle wiring diagram?

    a. 12 to 14 volts

    b. 0.25 to 1.25 volts

    c. 3.0 to 3.75 volts

    d. 4.0 to 4.8 volts

    Explanation: At a full throttle condition the TPS voltage should be close to the voltage reference of 5 volts. Answer d is correct.

14. A fuel pressure test is being performed on the composite vehicle. The fuel gauge measures 2 bar. This could be caused by all of the following EXCEPT:

    a. A faulty pressure regulator.

    b. A broken vacuum hose to the fuel pressure regulator.

    c. A restricted fuel filter.

    d. A restriction in the fuel inlet sock.

    Explanation: Two bar is equal to 29.4 psi. This is lower than specified on the composite vehicle. Answers a, c and d could cause lower than normal fuel pressure but a broken hose to the fuel pressure regulator (answer b) is the only thing which could not cause a lower fuel pressure. It would cause the pressure to go higher.

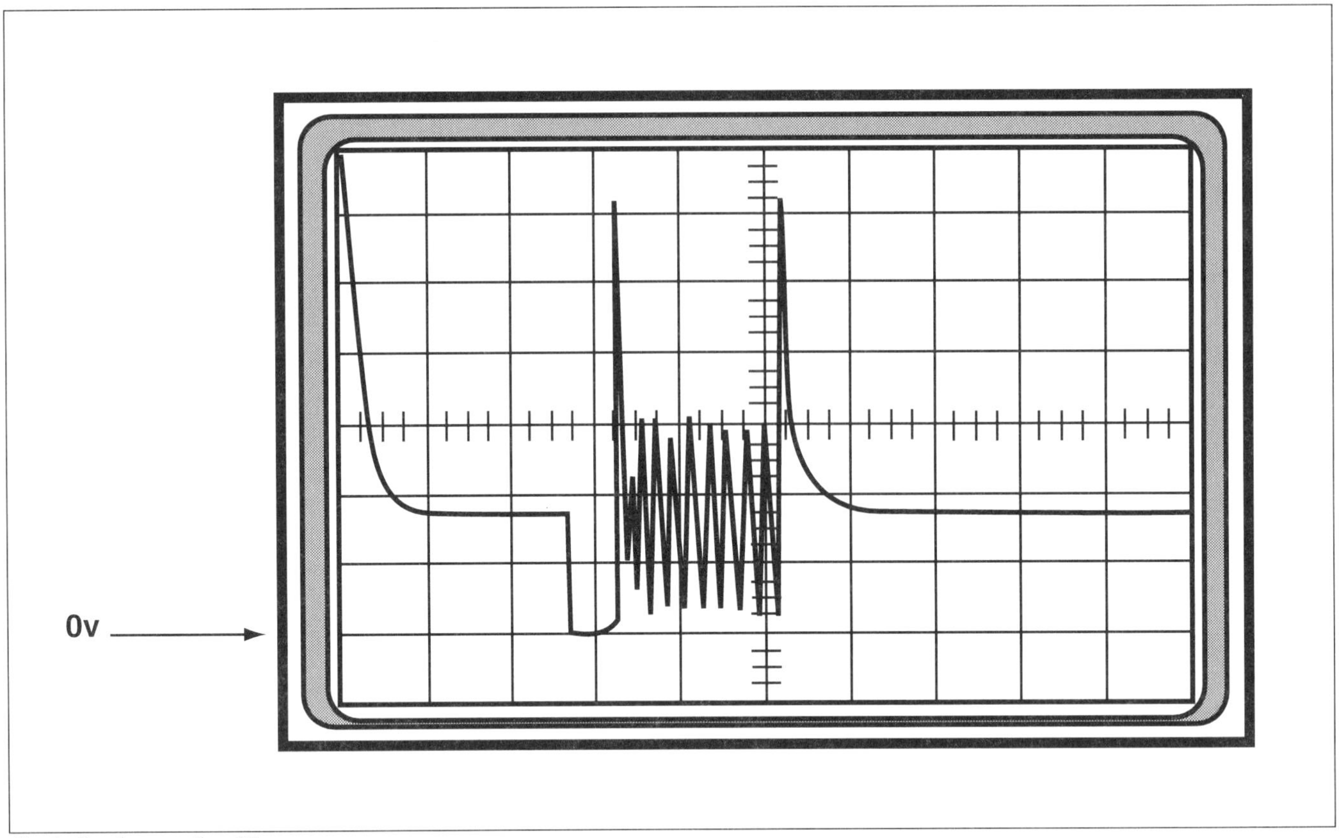

Illustration for question 15.

15. The TBI injector pattern above is seen taken at cruise.

    Technician A says this is normal current limiting injector pattern.

    Technician B says when current limiting takes place it is actually controlled by the PCM.

    Who is right?

    a. A only

    b. B only

    c. Both A and B

    d. Neither A nor B

    Explanation: Current limiting injectors modulate the voltage to control current. When the injector is first energized, it takes quite a bit of current to hold the solenoid open. Once it is turned on, only a small amount of current is required to keep the injector energized. When an injector is of the low resistance type, most computers modulate the current by varying the voltage to keep the PCM driver from overheating. Both technicians are correct.

16. Technician A says a piezoresistive knock sensor converts vibrating pressure as a result of engine knock into a varying resistance.

    Technician B says that most knock sensors are not piezoresistive but rather piezoelectric sensors.

    Who is right?

    a. A only

    b. B only

    c. Both A and B

    d. Neither A nor B

    Explanation: Most knock sensors are piezoelectric. However, there are also piezoresistive types. Both technicians are correct.

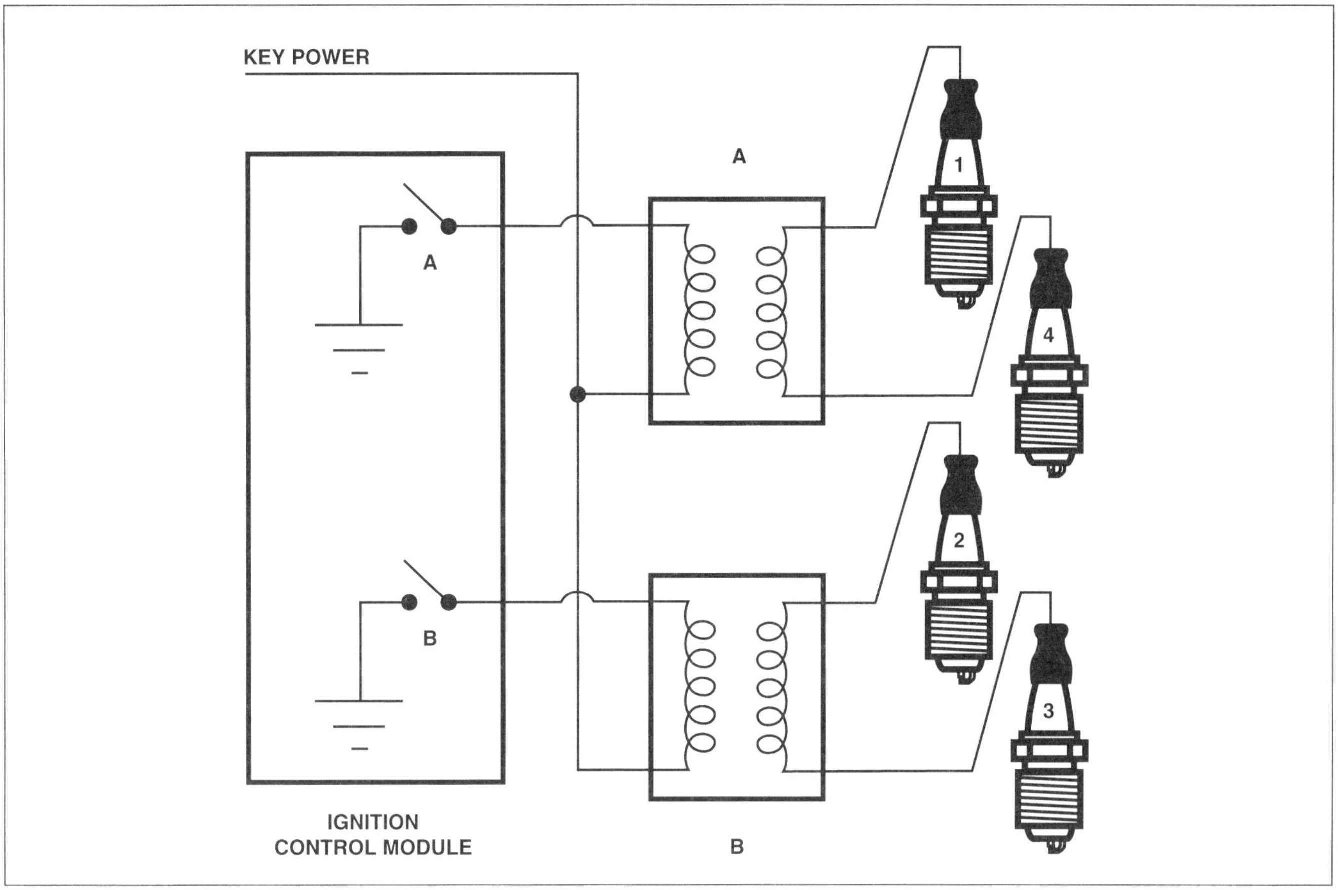

Illustration for question 18.

17. A voltage drop test is performed on the fuel pump circuit.

    Technician A says the voltage drop across the fuel pump is excessive; the fuel pump must be replaced.

    Technician B says the voltage drop across the fuel pump relay switch is excessive; the switch must be replaced.

    Who is right?

    a. A only

    b. B only

    c. Both A and B

    d. Neither A nor B

    Explanation: There should be very little voltage drop, if any, across the fuel pump relay switch. The maximum possible voltage drop across the fuel pump itself is desirable. Therefore, b is the correct answer.

18. Technician A says when ignition coil A fires, cylinders #1 and #4 fire virtually simultaneously.

    Technician B says one of those plugs will fire a positive voltage and the other plug at a negative voltage.

    Who is right?

    a. A only

    b. B only

    c. Both A and B

    d. Neither A nor B

    Explanation: Coil pack A fires both spark plugs #1 and #4 almost simultaneously. One spark plug will always fire with a forward current flow and its companion will always fire with a reserve current flow. Both technicians are correct.

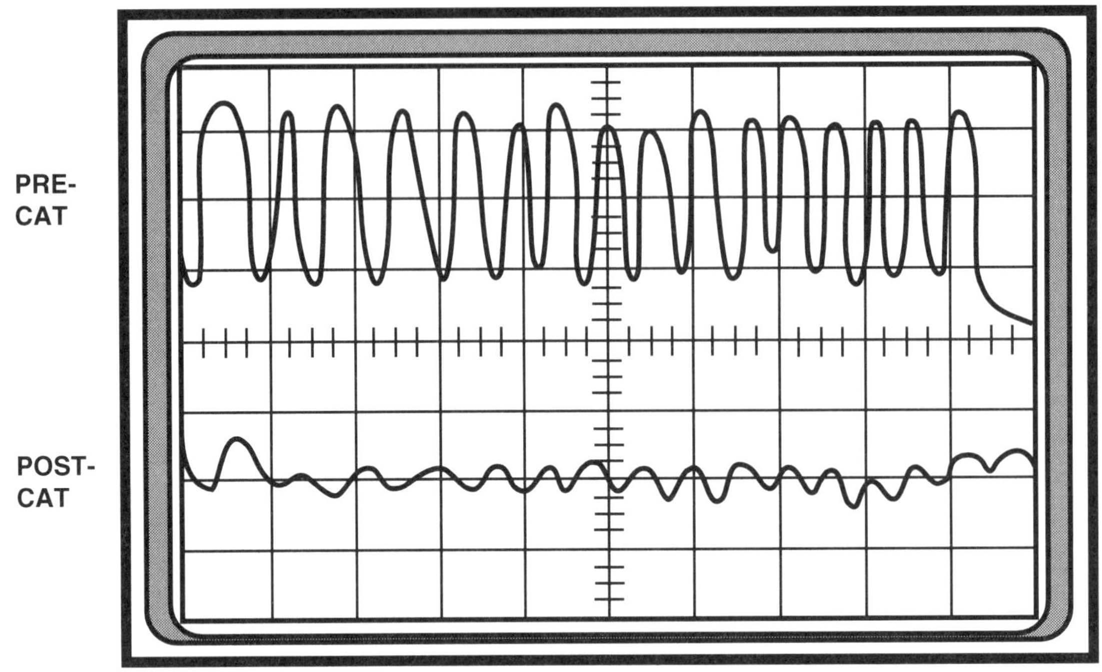

Illustration for question 19.

19. In the illustration above, an oscilloscope pattern for pre- and post-oxygen sensors are displayed; taken with the engine fully warmed up at 2000 rpm.

    Technician A says the pre-cat oxygen sensor is working properly. But the post-cat oxygen sensor shows very little response and should be replaced.

    Technician B says that both pre- and post-cat oxygen sensors are operating normally. There is nothing wrong with this picture.

    Who is right?

    a. A only
    b. B only
    c. Both A and B
    d. Neither A nor B

Explanation: The pre-cat oxygen sensor is operating normally. The post-cat oxygen sensor is also acting normally, so b is correct. If the post-cat oxygen sensor had a smooth oscillating pattern like the pre-cat oxygen sensor with a hot engine at cruise, a faulty catalytic converter would be indicated.

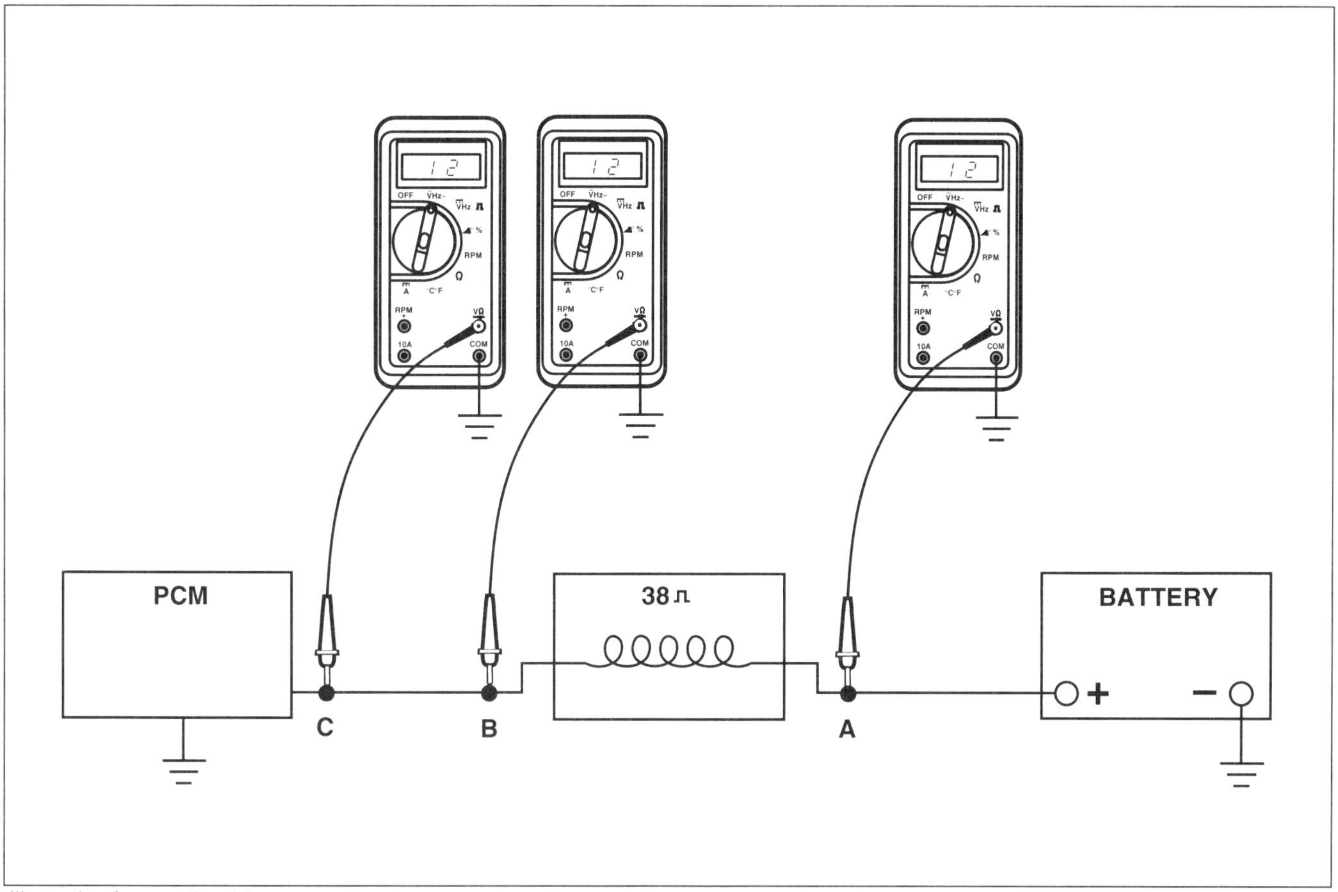

Illustration for question 20.

20. When checked with a voltmeter, points A, B, and C each displayed 12 volts.

    Technician A says this indicates that the solenoid is energized.

    Technician B says the voltage should be less than 1 volt at points B or C when the solenoid is energized.

Who is right?

a. A only

b. B only

c. Both A and B

d. Neither A nor B

Explanation: When a solenoid is energized by the PCM, the voltage will be pulled low when it is measured at either point B or C, so technician B is correct.